Isaac Pim Trimble

A Treatise on the Insect Enemies of Fruit and Fruit Trees

With Numerous Illustrations Drawn from Nature, by Hochstein, under the

Immediate Supervision of the Author

Isaac Pim Trimble

A Treatise on the Insect Enemies of Fruit and Fruit Trees
With Numerous Illustrations Drawn from Nature, by Hochstein, under the Immediate Supervision of the Author

ISBN/EAN: 9783337279882

Printed in Europe, USA, Canada, Australia, Japan

Cover: Foto ©berggeist007 / pixelio.de

More available books at **www.hansebooks.com**

TREATISE

ON THE

INSECT ENEMIES OF FRUIT AND FRUIT TREES.

WITH

Numerous Illustrations drawn from Nature, by Hochstein, under the immediate supervision of the Author.

BY

ISAAC P. TRIMBLE, M.D.,

ENTOMOLOGIST OF THE STATE AGRICULTURAL SOCIETY OF NEW JERSEY; ENTOMOLOGIST OF THE HORTICULTURAL ASSOCIATION OF THE AMERICAN INSTITUTE, ETC., ETC.

The Curculio and the Apple Moth.

NEW YORK:
WILLIAM WOOD & COMPANY, 61 WALKER STREET.
1865.

Entered according to Act of Congress, in the year 1865, by
WILLIAM WOOD & CO.,
In the Clerk's Office of the District Court of the United States for the Southern District of New York.

This Work

Has been prepared by one who has had an experience of many years in a vigorous and successful contest with the

INSECT ENEMIES OF FRUIT AND FRUIT TREES,

And is Dedicated to the

FARMERS AND FRUIT GROWERS OF OUR COUNTRY,

In the hope that it may assist in explaining some of the pages in the Great Book of Nature, in which their interests are deeply involved.

INTRODUCTION.

No other portion of the agriculture of our country is at this time receiving so much attention as that devoted to the culture of Fruit. The growing of young trees in nurseries, as an art, has been greatly improved, and the supply of such trees is enormous. Books have been prepared by men of eminent ability and mature experience, giving all necessary information for the management of the orchard; and the fruits that have been proved valuable have been carefully figured and described.

Some men are devoting much attention to the improvement of kinds, and with most satisfactory results. Fruits that were highly esteemed a few years ago are now superseded, and amateurs hope to identify their names with better sorts; and why not? There seems to be no limit set by nature to improvement in this direction. The Seckel Pear was found growing wild in a hedgerow. Probably the blossom that produced the pear, from the seed of which that tree grew, had been visited by a bee dusty with pollen from another blossom, and thus a germ was fertilized, which in time brought forth this hybrid of such surpassing excellence. The science of Botany teaches us that we can hybridize as well as bees, and improvements in fruits are now brought about as the designs of men rather than as the accidents of insects. Grafting, budding, and the propagation by cuttings give us the means of multiplying these better kinds with a rapidity characteristic of the age. But with all these signs of progress, the supply of fruit is far short of the wants of the people. The prices are often extravagant to the consumers, and do not always remunerate the producers.

There is no subject more frequently spoken of in Horticultural and Agricultural societies than the decay of fruit trees. We must all admit, that in the older States of our country, orchards do not flourish as they did fifty years ago, and the crops of every variety of fruit are becoming more uncertain. I have heard many discussions on this subject, and have often been surprised how little of the cause of this decay, or the uncertainty of the crop, is ever attributed to insect enemies. One person will ascribe all this change to exhaustion of soil; another to improper planting or defective cultivation. Others think there has been too little or too much pruning. Some will impute the defect to a want of the proper elements in the soil, or of a right proportion of those elements—either the lime, the potash, the clay, the sand, or the humus is not present, or not in the exact quantity to meet the demands of the growing tree or of the ripening fruit. I have heard farmers speak learnedly on this subject (quoting Liebig and other authorities), whose orchards were overrun with insect enemies that could have accounted for all their troubles, had they understood them.

Modern agriculture teaches the advantages of a rotation of crops, and it would be as unwise to plant an orchard where one of the same kind of fruit had stood before, as it would be to plant corn or sow wheat for a succession of seasons in the same field, unless it should be some alluvial spot of inexhaustible fertility. Most practical farmers know well that every soil can be exhausted by almost any crop under this improvident management. That has been the fate of large sections of this country. But proper rotation and more systematic manuring are changing all this. The soil is now *made* to produce paying crops, and can just as well be made to produce paying crops of fruit as anything else, if the trees and the fruits they bear are protected from the insect enemies. No farm will produce paying crops of wheat where the Hessian fly or wheat midge has taken possession. No prudent farmer will gather his crops year after year into barns infested with the weevil. We might as well suppose that the owner of a valuable flock of sheep that had been killed off by dogs, would expose another flock to a similar danger.

Fifty years ago the land in large sections of the State of New Jersey was considered "worn out." Whole counties were in a condition similar to that of the exhausted tobacco lands of Maryland and Virginia, but at that very time the State was famous for its crops of fruit. According to the census of 1860, the farming land of the State of New Jersey was worth about twenty dollars an acre more than the farming land of any other State in the Union. This is partly owing to its proximity to the markets of New York and Philadelphia, but chiefly to the great improvement in the productiveness of the soil by the use of marl and lime, two most valuable fertilizers found in great abundance. But the fruit crops of New Jersey have diminished in as great a ratio as the value of the lands has increased. This cannot be owing to the exhaustion of the soil. What, then, is the cause? In large sections of the State the Tent caterpillar is so numerous that the Apple trees are stripped of their leaves every year. Twenty and thirty nests are often seen on a single tree, and large orchards scarcely cast more shade than in winter. The leaves of trees are vital organs, the functions of which are similar to those of the lungs in animals. The Canker worms, Palmer worms, and several other species of caterpillars that feed upon the leaves of our fruit trees, are injurious just in proportion as they destroy these leaves. The owners of such orchards seldom disturb these caterpillars, and yet they complain of the premature decay of their trees, and tell you that raising Apples does not pay.

The Apple and Quince trees have no greater enemy than the Apple-tree Borer. One whose attention has never been called to the signs of the depredations of this insect will not suspect its existence till too late; while others who have investigated it carefully, will know its presence in an orchard by the appearance of the trees, even while passing them rapidly in a train of cars. This enemy is often brought in the young trees from the nursery. It is three years in coming to maturity, and increases slowly from such small beginnings. Young vigorous trees seem to resist for years, but as they begin to bear fruit the enemy increases faster than the growth of the tree, and the orchard dies.

The Peach worm feeds upon the inner bark, near the ground, each worm cutting off the connexion between the top and root of the tree to the extent of one or two inches. This insect is an annual; the next year's crop of worms will probably girdle that tree all round. The Peach-grower complains of the premature decay of his orchards, and says that peach trees are too short-lived to be profitable. Other cultivators understand this enemy, and "worm" their trees carefully, but will buy their stock from nurserymen who plant pits or use buds from trees diseased with the "Yellows." And they complain, too, of premature decay, and that a second crop of trees will not grow upon the same ground.

The Black Knot on Plum and Cherry trees is another increasing evil.

The Bark Louse or scale insect, found on both Apple and Pear trees, insignificant as it appears, often causes the speedy decay of orchards.

All the above insect enemies of fruit trees, as well as most of those of the fruits themselves, *are manageable—can be subjected to our control.* The man who permits them to increase and multiply, not only has no right to complain, but is a nuisance in his neighborhood, and should be treated as other nuisances are, that the public may be protected.

There are many other insect enemies quite serious at times, and not within the reach of our control, but most of them are transient evils. They are under the influence of checks wonderfully ordered for our protection. Some are brought to a speedy end by vicissitudes of weather. Birds come in flocks just at the right time for the destruction of others. Still more are subdued by insect parasites.

We import fruits of southern latitudes in large quantities. Do we export ours to a corresponding extent? *Can* we, when oranges in New York cost less than apples?

The preservation of fruits has long claimed much attention. They are wanted throughout the year. Men of wealth expend large sums in forcing-houses; artificial climates are thus created, that fresh fruits may be had out of season. People of moderate means dry and preserve them. Canning has lately been adopted to advantage, giving us in winter an approximating taste of the fruit luxuries of summer. Many attempts have been made to retard decay by the use of ice, but the fruits soon lose their natural flavor. There are said to be apples that will keep throughout the year. Many of the Russets will look like Apples till the next summer. A Northern Spy will still be good in April, but after that we must wait for Strawberries.

Most persevering efforts have been made for many years to obtain winter Pears. Some have been found that seemed to promise to be successful, but these have generally failed; and the best cultivators now have little expectation of being able to carry this superb fruit later than December.

I have lately heard Mr. Nice, of Cleveland, Ohio, explain his plan of constructing houses for keeping fruit, and have eaten freely of grapes and apples kept in one of these houses. The grapes had the same color, the same flavor, and almost the freshness of stem of well ripened Catawbas in their season. It seemed like eating them fresh from the vines in the spring instead of fall—April in place of October. A great variety of both fall and winter Apples were submitted for examination and tasting. Autumn apples were autumn apples still, and winter apples were not yet ripe.

Mr. N. says he has succeeded with all the fruits he has yet tried, except the Peach; with that he fails—probably from some influence of the down.

This manner of retarding decay appears to be founded upon correct principles of science, and has been perseveringly tested for several years. It seems to promise what we all want, good fresh fruit for every day in the year.

The preceding observations are intended to show—

1st. That the cultivation of fruit in our country is not so successful as it should be.

2d. That much more fruit is wanted than is likely to be raised, without some change of management.

3d. That the comparative failure of fruit is not owing to defects of soil.

4th. That all fruit trees will grow well in all parts of our country, on land that will produce good crops of anything else.

5th. That if successful plans of keeping fruit sound and fresh should be generally introduced, the quantity wanted will be greatly increased.

6th. That a foreign demand would be found for any superabundance.

As to the situation of orchards, no fruit trees should ever be planted on low wet ground. A western exposure, with protection from the morning sun, is best.

Trees whose buds are liable to be killed by the severe cold of winter, or the blossoms to come out early in the spring, should be planted in elevated situations. Plum trees grow best, and the fruit is generally finer, on clay soils, but light sandy lands are better for Peaches.

The health of your trees and your crops of fruit will depend upon how successful you are in subduing the Insect Enemies. If they are conquered, all who plant trees and manage them with reasonable care can have fruit.

As it has been the wish of the author to avoid the use of scientific terms, he has as far as possible confined himself to language that will be understood by all readers. But some explanations are necessary. The word Insect signifies *in sections*. Most insects, except spiders, are in three sections. They have six legs and two feelers or antennæ. The young of insects, as caterpillars and grubs, are often called worms, but improperly. Worms, or *vermes*, are never insects.

The science that treats of insects is called *Entomology*, and this, like other sciences, has the various species classified and arranged to facilitate their study. The primary division is into Orders. The insects spoken of in this volume belong to the two most important of these orders—the *Coleoptera*, or Beetles, and the *Lepidoptera*, or Butterflies. Nearly all the other enemies of Fruit and Fruit trees are included in these orders.

There are four stages in the lives of insects: the egg, larva, pupa, and imago. The word *larva* means *mask*. That is, the larva is a masked condition of the future butterfly. This word, *larva*, is commonly used to signify the embryo condition of insects generally; but in this work I have chosen to confine it exclusively to the Lepidoptera, and shall call the young of other Orders by other terms. Embryo beetles will be called grubs. The larva or caterpillar stage of the Lepidoptera, and the grub of the Coleoptera, is the period of their lives when they do the chief injury. *Pupa* means the *chrysalis* stage—the period of transformation from the embryo to the *imago* or image—the perfect insect.

The word Moth will often occur in a work like this, and may lead to confusion if not explained. The difference between a moth and a butterfly is, that the latter flies by daylight, the moth at night. In other words, the butterflies are the diurnal, the moths the nocturnal Lepidoptera. Butterflies and moths may be known from each other by the difference of the antennæ; the former having little knobs on the ends of their feelers, and the latter being without them.

This work is without plan as a scientific book. Although treating of insects, it does not arrange them into orders, classes, or families, but only discusses a few species, chiefly in the order of their importance as enemies of fruit and fruit trees.

The object of the Author has been to make a book to meet the wants of the practical man, who has but little time for the study of any subject except his business, and least of all, a science involving, as Entomology does, hundreds of thousands of species. To make such a work intelligible, *illustrations addressed to the eye are a necessity*. The fruit-grower should be able to identify his insect enemy positively when he sees it—there should be no guessing. The Curculio and Lady-bug, for instance, are both beetles; both are found upon the same trees; they will often fall down together when those trees are jarred. The one is our worst enemy, and the other one of our best friends. I have known people kill the friend and overlook the enemy.

I have been studying these enemies for many years. At first it was an investigation made necessary for the protection of my own crops; and that experience painfully taught me knowledge that I had not been able to find either in books or cabinets. The interest thus excited has been increased by the reading of such valuable works as those of Kirby and Spence, Huber, Latreille, Say, Harris, Fitch, and many others. From this reading and personal experience, I am satisfied that the interests of Fruit-growers would be promoted if all the practical knowledge on this subject could be gathered into a separate work, and I have felt that it was a duty to make a beginning by contributing my portion towards a better understanding of this difficult subject.

When I assert that any individual can subdue his fruit enemies if he chooses, I speak from my own positive knowledge; and although I do not wish to be understood to say that the instructions contained in this book are the best, I do wish to be understood to say, that some general plan of treatment should be adopted. An individual who resolutely determines to do it can save his fruits; but if all his neighbors for miles round shall act with him in carrying out the same instructions, the work of each will be less even the first year, and all subsequent seasons will be comparatively nothing. How such instructions are to be generally disseminated or such associations to be formed, it is not for an author to determine.

The next portion of this work, both the text and plates of which are in an advanced stage of preparation, will treat of the various Caterpillars injurious to Fruit trees and Grape vines. But the publication of an illustrated work like this is attended with so much expense that it is deemed advisable to await the verdict of the fruit-growing public before completing another part. If the public show by the reception of the present volume that more is wanted, both author and publisher will be encouraged to bring it out at an early period.

NEWARK, N. J., *April* 15, 1865.

THE CURCULIO.

PLATE I. (*Frontispiece.*)

1. The Promise.
2, 3, 4. The Fulfilment as it should be.
5. The Fulfilment as it is.
6. The Cause of the Difference.

THE Frontispiece of a book is generally intended to be looked at only. This one, if carefully studied, will convey an impression of the importance of one of the insect enemies that could hardly be realized by any arrangement of words.

Fig. 1 represents a cluster of the blossoms of the Apricot. This is the earliest of the fruit trees to bloom; the first evidence that Spring has really come. Few of the fruit blossoms are so beautiful as these, but like many others of the fair promises of this world, they are not always kept inviolate.

Fig. 2 is a Pond's Seedling Plum; see how large it is; look at the bloom; think how it would taste. The tree grows large and strong; bears full crops. Any one can have plenty of such plums if he will master the Curculio. There are more than fifty other kinds, all good, to choose from. The Plum tree is hardy; it will grow well in every part of our country, on every soil; no Borer at the root; no Yellows; no Curled Leaf. The buds are seldom killed with the severe cold of the winter, and the blossoms not often nipped with late frosts:

Look at Fig. 3, an Apricot—that superb fruit. Who eats an apricot? Look at it again, and resolve to have some. This fruit ripens during the last of July and early in August, between the Berries and the Peaches.

Fig. 4 is a Nectarine, the most beautiful of all the fruits. Many kinds are delicious. The tree is closely allied to the Peach, growing where it will. Get it clear of the taint of the Yellows, and keep out the Borers, and the Curculio is the only enemy to prevent your having the fruit in abundance. Think of such ornaments in your garden, such embellishments to your table, such presents for your friends, *and be cowards no longer.* We have but one life to live in this world; let us enjoy it; let us feast our eyes on such beauties; let us indulge our appetites on such luxuries—they are innocent. We were created with dominion, and have a right to subdue the Curculio, the only enemy that stands between these fruits and us. It will be hard to conquer, undoubtedly, but we can do it. Then let us all join hands, and pledge ourselves to do both, and thus try to make the world what it should be.

Fig. 5 represents the work of the Curculio.

Fig. 6 is the Curculio itself.

PLATE II

PLATE II.*

1. Pear, punctured May 23, 1863.
2. Curculio cutting the crescent.
3. The position of the Curculio when making the cavity for the reception of the egg.
4. Plum. The first mark, May 25, 1863.
5. Cherry. The first mark, May 25, 1863.
6. Peach. The first mark, May 28, 1863.
7. Apple. The first mark, June 1, 1863.
8. Apple with two perfect punctures, and others imperfect.
9. Apple. From a bottle containing a colony of Curculios.
10. Siberian Crab-Apple. Marked June 10, 1863.

That part of the season between May 18th and June 10th, the period included in the illustrations on this Plate, is an important time to the fruit-grower who has determined to save his crops from the Curculio.

All kinds of Pears and Cherries will not be large enough for the Curculio's operations at these dates, and most of the Plums will be a day or two later.

Apricots will generally be from a week to ten days earlier than any other fruit, and this crop will often be attacked by the Curculio while other kinds of fruit trees are still in blossom.

Occasionally there will be a season when the blossoms on nearly all fruit trees will burst together; the Apricot, Pear, Plum, Cherry, Nectarine, and Peach, presenting their beautiful promises at the same time.

In this case, the young fruits will come so nearly together as to give the Curculio its choice, and the Nectarine will be chosen. The reason why the Apricot is so generally destroyed by the Curculio, is probably owing to the fact of its being, for several days, the only fruit large enough for its use.

---

* The original paintings of the illustrations of the Curculio were made in the years 1863 and 1864, at Newark, New Jersey, in the latitude of 40° 45'. By a reference to the diaries of these two years, there appears a difference of five days in the time of the first punctures by the Curculio in the same fruits. In 1863 it was May 23d, in 1864 it was May 18th.

If all the fruits were of the proper size at the same time, they might be placed in the following order as to their liability to be attacked by the Curculio: Nectarine, Plum, Apricot, Apple, Pear, Quince. Some varieties of the different kinds are preferred to others. The Green Gage, Washington, and Egg Plum will suffer more than the Prunes, Damsons, and many of the common kinds. The earliest Apples, as the Sweet Bough and Early Harvest, will be more injured than later kinds.

The black knot, so often found on Plum and Cherry trees, is used freely by the Curculio. These knots are often several days in advance of the young fruit, and the female Curculio has been known to exhaust her supply of eggs in these knots before the young cherries or plums on the same trees were fully formed.

Figure 2 shows the position of the Curculio when cutting the semicircle or crescent-shaped mark. This is made by the end of the proboscis, and merely goes through the skin. This part of the process, while the fruit is young and tender, is soon finished, sometimes not taking more than two or three minutes.

Fig. 3 shows her position in the next part of the work. From the centre of the concave part of the crescent, the proboscis is introduced under this cut skin, and there it slowly works, cutting its way until it can reach no further.

The end of this cell or cavity is now dug out or enlarged, to make it a suitable receptacle for the destined egg. The insect has an instinct which teaches her that the surroundings of this cavity must be so deadened that no subsequent growth of the fruit at this part shall press upon that delicate egg and crush it. The seventeen-year Locust arranges her eggs crosswise in cells made in the twigs of growing wood; but on one side of each cell the wood is so comminuted by the boring instrument of the female Locust that it never recovers; and although the twig generally continues to grow, this wounded part will not be grown over until long after the eggs have hatched. Were it not for this instinctive foresight of the necessity of so splintering up the wood on a side of the cavity where one end of these oblong eggs rests, that it would yield to the pressure from the other, in the growth of two months, these eggs must be broken. The Curculio probably has a similar instinctive foresight.

The preparation of this cell is much the most tedious part of the process, usually taking about fifteen minutes, though sometimes half an hour. During most of this time the Curculio will be found in this pitching position, and with her proboscis entirely buried; looking as the woodcock does when boring for food in the soft ground. This cavity finished, she turns round and deposits an egg at its orifice; then assuming the former position, very quietly pushes that egg with her proboscis to its destined place. Next, the crescent-shaped cut is plastered up with a gummy substance that holds the cut edges together for the time being; probably an instinctive precaution against the weather or insect enemies that might endanger the safety of that egg. The female Pea-Bug deposits her egg in a slight wound in the pea-pod, and then covers it over with a tenacious paste.

Fig. 8 is intended to show an Apple with two Curculio marks perfected, and several others partly finished. Some writers have said that the Curculio never deposits more than one egg in a fruit; but this is a mistake. Two or more grubs will often be found; but these instances of so many marks so early in the season are rare except in the apple, and, if examined, most of them will be found unfinished and containing no eggs.

Those parasitic insects that introduce their eggs into the living bodies of other insects, graduate the number so deposited according to the quantity of food the body of that insect will afford the young from those eggs, so that each shall have enough to bring it to full maturity. The instinct that teaches this knowledge is unerring. A large Ichneumon Fly will not deposit an egg in a caterpillar too small to afford the requisite amount of food, but she will select one that will just yield enough, and none to spare. If the caterpillar is of such a size as to feed two, three, or more, two, three, or more eggs will be deposited in it.

The Curculio probably has a similar instinct. A young apple or peach, dry and withered, and not larger than a hazel-nut, will be found containing the grub of a Curculio full grown, plump, and active, with no part of that apple or peach left that could be used for fruit; but it would be hard to find two half-grown grubs in such a fruit, and with nothing more to eat.

Fig. 9 is an Apple that was left some time in a bottle with a colony of Curculios, and the number of punctures shows that instinct ceases to be a guide under such circumstances. This apple was one of three that had been coated over with the "whale-oil soap" mixture, which, with three other apples taken fresh from the tree, was given to these Curculios at the same time, to test whether they would make any distinction. But none could be observed. All were taken with the same avidity, and each was equally punctured.

Fig. 10 shows that even the Crab Apple does not escape the Curculio. I have seen the unmistakable crescent-shaped mark of this insect on that most minute of apples, the Currant Crab.

Figs. 4, 5, 6, 7, and 10, of this Plate give good representations of the mark of the Curculio, where the entire process has been completed, and the egg has been secured. When first made it is not so distinct, but it soon becomes discolored, presenting this brown appearance. There will often be a slight convex elevation over the egg, and a gentle pressure on that spot with the thumb-nail will break it. A very sharp sense of hearing will sometimes detect the snap.

UNIV. OF
CALIFORNIA

PLATE III

PLATE III.

1. Plum, June 3. Egg of Curculio hatched, and the young grub eating its way.
2. Same Plum, June 3. A slice cut off, exposing the passage-way of the grub.
    a. The grub, the natural size at this date.
3. Green Gage, June 14th. Several Curculio marks. The globules of gum indicate that the eggs have been hatched.
4. Peach, June 14. The particles of gum here indicate the same thing as in the Plum, Figure 3.
5. Washington or Bolmar Plum, June 20th. Fruit fallen, and the grub emerged from hole at a.
6. Peach, June 24. On the ground, a grub having just escaped.
7. Pear, June 24. Showing blemishes from punctures of the Curculio.
8. Perfect Cherry, June 25.
9. External appearance of Cherry containing a grub of the Curculio when nearly full grown.
10. The same Cherry when opened.
9 and 10. The kinds of Cherries which birds prefer.

THE dark line leading from the puncture in Fig. 1, is a common appearance in the progress of the Curculio, but not universal. It indicates unmistakably the destruction of the Plum.

Fig. 2 shows that the young grub does not proceed at once towards the centre of the plum, but soon after this it will be found feeding round, or in the pit itself.

The globules of gum on Figs. 3 and 4 are proof positive that nothing can save such fruits. Apples also sometimes show gummy exudations from the wounds made by the Curculio, but this gum does not become so concreted as in the plum and peach, remaining soft and sticky.

Peach-growers will recognise Fig. 6 as a kind of peach too often met with under their trees about the last of June and the early part of July.

Fig. 7 represents a very common appearance of the pear.

In a plantation of Pear trees standing by the side of an old neglected Apple orchard, I have caught several hundred Curculios in less than an hour, by jarring thirty or forty trees. But as soon as the neighboring apples were large enough the pears would be deserted. The Pear, though often injured, suffers less from this

PLATE IV

PLATE IV.

1. The Plum, June 9, showing the appearance when the egg of the Curculio has been taken out.
2. The German Prune (Quetsche), July 18, with two Curculio marks on the neck, dried up.
3. The Fellenberg Prune, ripe September 15.
4. The Green Gage as it should be, September 1.
5. The Green Gage, July 20, showing the origin of the rot.
6. Its further progress, August 7.
7. Nectar, in New Jersey, September 1. Rochester, September 15.

This Plate indicates the condition of some of the Plums later in the season. Fig. 1 shows a wound made by taking out the egg of the Curculio with the nail of the little finger. This is, however, too large an instrument for so delicate an operation, and leaves an unnecessary scar. Apricots, Nectarines, and Plums can be saved from destruction in this manner.

In the early part of the season, especially if the weather should be cloudy and cold, it will often be a week or ten days before this egg hatches; but in very hot weather the young grub will escape in four or five days. All attempts to save the fruit after the egg is hatched will be useless.

Those who have young fruit of valuable sorts not yet tested by the tasting process, will be anxious that the first crop shall come to maturity; and to know what to do when all have been punctured by the Curculio, will be useful information. The best instrument I have found for this delicate operation is a common quill tooth-pick slightly rounded at the point, and pared to a cutting edge. This must be insinuated under the concave side of the crescent-shaped mark, so as to turn over the triangular portion of skin lying between the horns of the crescent and the end of the tube where the egg is deposited. The egg—a white round speck—will sometimes be exposed, and a very sharp eye will detect it without the assistance of a glass; but generally it will be so coated with a covering of the pulp of the fruit as to be invisible. Take off this speck of skin, egg and all. If properly done the fruit will come to maturity, showing scarcely a blemish.

I have often carried such patients as these safely through a number of these otherwise fatal wounds. Retired gentlemen of leisure, who become amateur fruit-growers, can find amusement in this manner of fighting the Turk; and although rather a tedious operation, it is much better than cutting down the trees, or swearing at the Curculio as some do.

Figs. 2 and 3 are German Prunes, which are now becoming more popular than other Plums, from an impression that they are less liable to be attacked by the Curculio.

Fig. 2 represents a Prune with two Curculio marks on the neck, both dried up and harmless. This is a very common appearance of prunes late in the season. Why the Curculio so generally chooses the neck of this class of plums, or why the egg so often fails, I have not seen satisfactorily accounted for. This is an inducement for cultivating the Prunes in preference to other Plums, to those who do not intend to *conquer* the Curculio.

Fig. 3 is the Fellenberg Prune, ripe the middle of September. Some others are superior in flavor, and some ripen so late in the fall as to be more valuable on that account. When the Curculio shall be disposed of, and plums recovered from among the lost good things, many of the prunes will become favorite fruits.

Fig. 4, in this Plate, is the Green Gage—the good old Reine Claude.

Figs. 5 and 6 show what it becomes under the management of the Curculio.

Fig. 7. The same, when ready for the palate; and the palate probably never receives a pleasanter sensation. Had our Mother Eve been tempted with such fruit instead of apples, when she "brought sin into the world, and all our woe," she would have been more excusable.

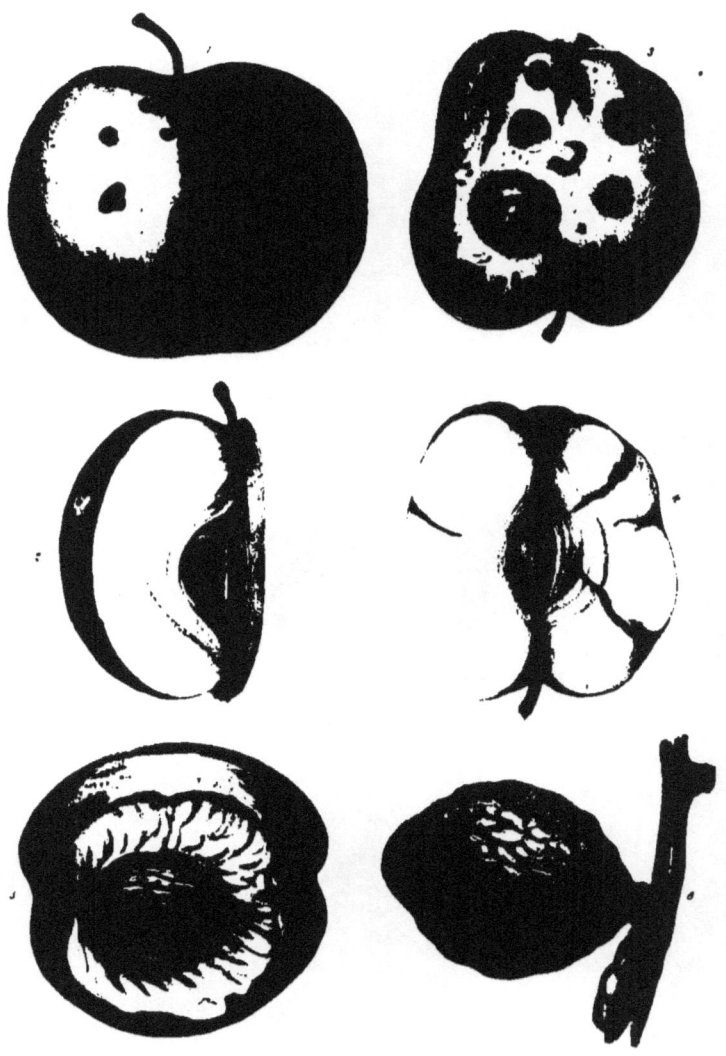
PLATE V.

PLATE V.

1. The Maiden's Blush Apple, Oct. 22, about half size, showing several Curculio marks, the largest in shape of a shield.
2. A section of the same, cut through some of these marks, and showing no injury under the skin.
3. An Apple, Oct. 22, greatly deformed by the punctures of the Curculio.
4. A section of Fig. 3. The four dark lines indicate the course of the young grubs of the Curculio.
5. An early Peach, Aug. 5. The first bite.
6. Crawford's Late, Oct. 4th.

THIS Plate shows the progress of the Curculio's operations upon Apples and Peaches.

Figs. 1 and 2 are intended to show that many of the punctures made by the Curculio upon apples do no serious injury—merely leaving blemishes only skin deep. These scars are of many forms, but the most common is in the shape of a shield, like the largest mark in Fig. 1. This shield varies greatly in size and shape. If the puncture has been made on a large kind of apple very early in the season, it will often expand with the growth of the fruit till it becomes two or three times the size of the one in the Figure. Generally the shield-shaped mark will be wider and shorter than here represented.

In nearly all of these wounds, of whatever shape, there may be seen a little spot darker colored than other parts, indicating where the egg was originally deposited. More or less of these marks can be seen on most of even the best Apples in the markets of New York and Philadelphia, both summer and winter, no matter from what part of the country they may have come.

Fig. 3 represents what is often called the "*gnarly*" fruit. This deformity is caused chiefly by the Curculio. It is very often seen in the orchard, and is especially common at cider mills.

Fig. 4 is a section of the same, showing the passage-ways of the young grubs towards the centre of the apple. When one of these roads is cut through the centre, it will look very much as represented in this Figure; but if the knife should pass a little to one side of the centre, it will be of a green or yellow-green color, and much narrower, looking as if it had acted as a cord, tying the centre and circumference of the apple together, and preventing its expansion in that direction, and thus causing the depression. Apples, when wounded in this way, are not destroyed, only deformed. They usually hang on the trees to maturity. The grub, from some cause or other, perishes before reaching the vital part. This was the condition of the apple crop in large sections of the country in 1864. There had been a period of nearly six weeks in July and August of excessively hot and dry weather, after which very few living grubs of the Curculio could be found. The marks, as seen in this Figure, where they had gone just so far, were very numerous.

We sometimes encounter the grub of the Curculio in early peaches, as we do in cherries, apricots, and early plums.

Fig. 6 was taken from a tree of the Crawford's Late at the time the sound fruit was ripe. The puncture of the Curculio had caused it to rot. Gum had exuded from near the stem, sticking it fast to the twig. Such specimens of fruit, still more dried up and withered, may be seen on Plum, Peach, Nectarine, and Apricot trees, often hanging on all winter.

Previous to the Rebellion, cherries, apricots, early apples, and peaches, were brought to the New York market from many of the Southern States, often from as far south as Georgia. If there had been no other evidence that the Curculio was common in that section of country, these fruits would have settled the question. Terrible as this pest is with us and further north, when the same fruits from the different sections were subjected to a comparison, the North would seem to suffer least.

PLATE VI

PLATE VI.

1. Curculio, natural size.
2. Curculio, with wings expanded.
3. Portion of ??? July 16, with the cells made by the ??? of the Curculio, in which they are undergoing the ???
4. ???
5. ???
6. ???
7. The ???
8. ???
9. ???
10. A species of Curculio often seen in September and October. This was figured from a specimen taken from the stomach of a Toad.

The most systematic treatise on American Entomology is that of Thomas Say. It is purely scientific, and to those who study insects as a science such works as this ??? a necessity. But little progress can be made in a field of investigation so ??? without systematic classification. Entomologists will speak of Orders, ???, Families, Genera, and Species, as ??? and ??? do. The structure of a ??? would be a ??? were it not for such a fixed language.

The following is Say's account of the Curculio. I give it in the exact words, as taken from page 285, vol. i. New York edition of 1859.

"7. C. ARGULA Fabr. (Rhynchænus) Syst. Eleut.
"Curculio nenuphar Herbst. Natursyst.
"R. cerasi Peck, Jour. Mass. Agr. Soc., Jany. 1819.
"This also varies much in size, and depredates on the plumb and peach and ??? fruits. My kinsman the late excellent Wm. Bartram, informed me that it also destroys the European Walnut in this country."

??? but a vague idea to the farmer or fruit-grower who has paid no

PLATE VI

PLATE VI.

1. Curculio, natural size.
2. Curculio, with wings expanded.
3. Portion of earth, July 16, with the cells made by the grubs of the Curculio, in which they are undergoing their transformations.
4, 5. Front and back views of the pupa of the Curculio, July 16, greatly enlarged.
6. The Curculio more enlarged, almost matured, and just ready to emerge from the ground. It is now of a reddish color.
7. The Curculio still more enlarged, with the wings, legs, proboscis, and antennæ expanded.
8. Shows the proboscis, antennæ, and lenses of the eye.
9. Pea-Bug, twice the natural size.
10. A species of Curculio often seen in September and October. This was figured from a specimen taken from the stomach of a Toad.

THE most systematic treatise on American Entomology is that of Thomas Say. It is purely scientific, and to those who study insects as a science such works as this become a necessity. But little progress can be made in a field of investigation so immense without systematic classification. Entomologists will speak of Orders, Classes, Families, Genera, and Species, as ornithologists and botanists do. The structure of a science would be a Babel were it not for such a fixed language.

The following is Say's account of the Curculio. I give it in the exact words, as taken from page 285, vol. i. New York edition of 1859.

"7. C. ARGULA Fabr. (Rhynchænus) Syst. Eleut.
"Curculio nenuphar Herbst. Natursyst.
"R. cerasi Peck, Jour. Mass. Agr. Soc., Jany. 1819.
"This also varies much in size, and depredates on the plumb and peach and other stone fruits. My kinsman, the late excellent Wm. Bartram, informed me that it also destroys the European Walnut in this country."

This conveys but a vague idea to the farmer or fruit-grower who has paid no attention to the science, and will certainly not give sufficient information to enable him to save his fruit crops from this terrible enemy. My description is addressed to the eye. Look at the first figure on this Plate. Examine it carefully. That is the Curculio—the Plum Weevil—the Turk—the Little Joker, that I am making such a fuss about.

The sad effects of the Curculio upon the fruits, as shown in the preceding Plates, will prepare the reader to understand what a terrible evil it is, and he will be likely to study this Plate with greater interest than if it had been the first of the series illustrating this insect.

Of the four hundred thousand species of insects known to naturalists, the Curculio or Plum Weevil is the most important. Other insects are often more destructive for a time, but their ravages are transient, most of them being brought to a sudden end by natural causes. The Curculio has increased constantly since it was first noticed by fruit-growers, during the latter half of the last century, and is now found in nearly all the settled parts of North America, except the States on the Pacific Coast. It spreads with the settlement of the country, and increases with the multiplication of fruit trees. It has never yet been controlled in an appreciable degree by human agencies. Parasitic enemies cannot reach it. Vicissitudes of weather, except in localities and for a short time, have never checked it. It is marching on, "conquering and to conquer," unless there shall be concerted intelligence, and concerted effort to stop it.

Almost every person who owns a fruit-tree suffers more or less from this insect enemy. The fruit-growers lose a part of every crop every year, and the fruit-consumers get less than half that the same money would buy, if it were not for the Curculio. As a liberal supply of fruit adds greatly to the health and comfort of the people in all countries, but especially in a climate so heated as ours, the loss thus sustained is a serious matter.

We have no data upon which it would be possible to calculate the amount of damage caused every year by this insidious enemy, but we may safely estimate it by millions of dollars. A single living Curculio weighs a quarter of a grain, and it therefore takes about twenty-eight thousand to make a pound. If we take three quarter-ounce vials, and put 100 Curculios in one, 100 Pea-Bugs in another, and 100 grains of buckwheat in the third, each will appear about half full, and they will all look so much alike in size and color, that at a short distance they cannot be distinguished from each other.

Many people think insects too small to be worthy of much attention. Such people should consider the single grain of wheat, or the individual rain-drop. The Coral insect, in the abstract, is wonderfully insignificant, but the Coral insect in the concrete changes the channels and currents of the ocean, and builds up islands from the sea.

Fig. 1, in this Plate, represents the Curculio, as nearly correct in size, form, and color, as can be made in a drawing on stone. The antennæ are rather too heavy, but it is difficult to get the exact size of an object so minute; and the chest appears too deep, but a close examination will show that this apparent disproportion is owing to the position of the upper part of the fore leg. With these corrections understood, any one will be able to identify the living Curculio by comparing it with this figure.

Fig. 2 does not represent the Curculio as seen flying—that would be difficult—but as a dead one appears with the wings and legs spread out.

Fig. 3 shows a portion of earth with the young Curculios in their cells, undergoing their transformations. This piece of earth was taken from the centre of a flower-pot that had been filled two-thirds full of common garden mould, and on which the punctured plums that had fallen from a Green Gage tree had been thrown every day as they fell. The number of cells will be in proportion to the number of Plums thrown into the flower-pot. I have seen such earth almost as cellular as a honey-comb.

The effect of drought can be tested readily by experiments of this kind. Place one such flower-pot in a building, and throw water on it occasionally, as it would be rained upon out-of-doors, and let another remain perfectly dry. The grubs in the first will come out beetles; in the other they will perish apparently for want of moisture. This fact will have an important bearing when we come to consider the effects of the vicissitudes of the weather, not only upon this but upon many other insects which have a great influence upon human affairs.

The Plum crop fails for a series of years, and then for a single season will be abundant. I have raised full crops of Nectarines, Apricots, and Plums, every year for ten years in succession; but all those crops, except one, were the result of most persistent fighting the Curculio. The year of that one exception had been preceded by a local drought. For several weeks during July and August it had not rained in that neighborhood. Showers were often threatened, so that farmers hurried to secure their hay and grain; but the rains did not come. The earth became as dry and parched as if it had been in flower-pots and under cover.

We often complain of the weather. The severe cold pinches us, and is hard to bear, especially as we outgrow the love of skating or sleigh-riding. Excessive heat is equally uncomfortable. Rains, to some, never seem to come exactly at the right time. Crops will be injured in harvest; pleasure parties will be broken up. Long continued droughts, with a brazen sun setting day after day, unmistakably indicating no rain to-morrow, make us feel how powerless we are to avert impending famine. What was planted gives no increase. The pastures fail, "the Grasshopper becomes a burthen," and we complain. But since that season's exemption from the Curculio I have learned to be more patient. These rough extremes have their compensations. I have known a terrible raid of Mosquitoes ended by a cold night in June. The papers in some of the Western States told us that one of the insect enemies of the wheat crop was killed by that same cold night. The Chinch bug goes on increasing, and its ravages become more and more serious until arrested by a rain-storm. The Aphides, that sometimes fairly blacken the young shoots of grape vines, will be melted into mere stains by a single shower. Wasps kill their young after the first frosty night. No insects are more attentive to their broods than the wasps, but they seem to know from instinct that the days of these young will be "few and evil," after such a warning.

For years a species of Thrips had been living on some Sugar Maple trees near my house They had become so numerous that every leaf had its colony, and the foliage had turned grey by reason of their sapping operations. On the 25th of June the mercury in the thermometer rose to 100° in the shade, and next day not a living Thrips could be found on those Maple trees. A slight breeze drifted them in eddies upon the piazza, looking like the seeds and chaff of timothy on a barn floor.

Almost every one will remember an occasional crop of Plums coming to maturity. I have heard of many such instances; and where there has been a chance to investigate have found that they have been preceded by a summer drought the year before.

The cells in Figure 3 are usually made from three to six inches below the surface, and the grub shapes them by a succession of turnings and twistings as a bird forms her nest, or as a dog will prepare his bed. The grubs of many beetles that go through their transformations in the ground do just so.

Figs. 4 and 5 are greatly enlarged representations, both front and back, of the pupa, giving very satisfactory views of the appearance of this insect in its intermediate stage between the grub and the beetle. The Curculio in this stage has no power of locomotion, but it shows its sensibility when the cell is broken into by a restless, wriggling motion.

Fig. 6 is a representation of the matured beetle before it has emerged from the ground. It will be found of various shades of color, but generally of a pinkish red; these colors, however, soon change into those of Figure 1 of this Plate, after the insect comes to the surface.

Fig. 7 is only a greatly enlarged view of Fig. 2, to make it more satisfactory.

Fig. 8 gives a good view of the origin of the antennæ or feelers.

The eyes of most insects are wonderfully formed. They may be said to be compound eyes, each made up of many hexagonal lenses. If a comb of the hive bee, containing one or two hundred cells, could be photographed down to the size of the head of a pin, it would look somewhat like the eye of a beetle. Each eye of the Curculio contains about 150 of these lenses. The number in the eyes of Butterflies, Moths, or Dragonflies, amounts to many thousands. In some microscopic experiments made last summer upon the eyes of plant lice from different trees and plants, it was found that the number of lenses in the eyes of these insects varied from every tree and plant. Each thus proved to be a distinct species, no matter how close the resemblance in other respects. Thus, should the rose bushes of a garden or a neighborhood be cleared of these pests they would not be re-inhabited by those from other plants. While examining one of these aphides it brought forth a young one, and this in turn being tested its eye was found to contain the same number of lenses as the mother's. This peculiarity of the eyes of insects, and the knowledge of the exact number of these lenses in the eye of each species, become important in investigations where only the comminuted parts can be obtained. In a long series of examinations of the contents of the stomachs of birds, for the purpose of ascertaining more positively how far the insectivorous kinds frequenting orchards are useful in feeding upon these enemies of the fruits, the microscope has enabled me to demonstrate many facts otherwise difficult to prove.

Figure 8 of this Plate shows the arrangement of the lenses of the eye. An ordinary pocket-glass will reveal this in many of the beetles, some of the large flies, and many other insects; but so minute are these lenses in the eyes of Butterflies, Moths, Dragonflies, etc., that a glass of much higher power will be required.

UNIV. OF
CALIFORNIA

PLATE VII.

PLATE VII.

1. The Canvas.
2. Long Stretcher.
3. One of the Short Stretchers.
4. Slip to Straddle the Tree.
5. The Mop Stick padded for jarring the branches of large trees.
6. The place where a branch has been cut off, leaving an inch or two of stump for striking upon with a mallet.
7. Plum with a Curculio as it appears after the first blow or jar—having withdrawn her proboscis and doubled up her limbs ready for the fall to the ground when the jar is repeated.
8 & 9. The Curculios and dead Plum buds—showing their resemblance.

This Plate is an illustration of the only effectual way of managing the Curculio when it comes upon the fruit. "An ounce of prevention is said to be worth a pound of cure." But if there has been no effort at prevention, the canvas, mallet, and mopstick become a necessity. This manner of mastering the Curculio involves much labor. Few will undertake it, and many who do will not persevere to the end.

My plans of fighting the Curculio are few and simple. Destroy all in the embryo condition, if possible. Every fruit, whether nectarine, apricot, plum, apple, pear, or quince, containing the grub of the future Curculio, falls prematurely from the tree. This grub remains in that fallen fruit long enough to give plenty of time for its destruction. All our domestic animals, horses, cattle, hogs, and sheep, will eat these fruits if they have the chance. Poultry are also recommended, but are not to be depended on except for cherries. Where it is impracticable to use animals for this purpose, let all these young fruits be gathered by hand as soon as possible after they fall, and then destroyed. They may be fed to the stock or burnt. Let there be no exceptions on the whole farm. Some Apple or Cherry tree may stand in an out-of-the-way place, an unsuspected breeder of this pest for years. If the fruit on such a tree is not valuable enough to have it attended to in this way cut it down at once. By all means cut down all useless or superfluous Cherry trees, and see that the remaining trees of this fruit stand where the hogs and poultry have free access. Form neighborhood associations—fruit-growers' clubs, where all shall do the same thing. Do this faithfully a single year, and the benefit will be so apparent, in more and better fruit, that it will be done the next year as a matter of course, and every succeeding year the labor will be less and the benefit greater. Don't stop because one surly fellow will not join you, or because other neighborhoods will not do it. If there should be but one such fruit-growers' club in a county or state, the members will have the more labor, of course, but there will be the greater profit. If your neighbors will not join you, then fight the battle alone. Show them it can be done; let them see the fruits, and shame them.

Plant Plum, Apricot, and Nectarine trees—plant orchards of Apple, Pear, and Peach trees. Have fruits so plenty, and of such valuable sorts, as not only to pay the expenses of the extra labor but leave a handsome profit, after using all you want of the best for yourself and family. If you have not destroyed all the Curculios when grubs, or if your neighbors have not joined you, and they come upon your young fruits, then *at them with the canvas. If this is properly managed your fruit can be brought to full maturity as certainly as if there were no Curculio.*

If your trees are young—the first, second, or third crop—a canvas six feet square will answer well, and you can manage it alone. The palm, or rather the heel of the hand, will do the jarring. Some of the Turks will come down with the first blow, more with the second, and but few after the third. This bringing down the Curculio is to be done with a blow—*a sudden jar—not a shake*. Though the wind shakes a tree, the Curculio does not stop work on that account, but a jar alarms her instantly.

Fig. 7 of this Plate indicates the position of the Curculio on a Plum under the alarm of the first blow; her proboscis is withdrawn at once, the claws cease to hold fast, the limbs are drawn up, and the fore legs doubled at the knees, and one placed on each side of the proboscis. Another jar, and she falls to the ground, and there, among the grass and dead buds, as seen at Fig. 8, feigns death—an instinct of self-preservation common to insect life. Children find collections of brilliant little beetles, looking like beads, and they string them with needle and thread. The beetles cannot " play possum " long after such an operation, and the beads are found to have legs. The travelling lady-bug stops if you touch her. A little " bouncing beetle" will throw itself to a distance if you attempt to catch it, and there lie perfectly still. Kirby says the cockchafer will feign death if it sees the approaching rook, and has not time to secrete itself, not, as the Curculio does, by drawing itself up into a round ball, but will spread its legs out at full length, and look as a dead cockchafer should, knowing that a rook will not eat a bug that he does not kill, and that this sprawling position is a sign that it is already dead. This may be so, but I do not take the responsibility. We have no rooks.

Caterpillars are sought after as food by birds, just in proportion as they are clear of hairs. The *Geometers*, as the Canker worms and Span worms, are of this kind. I often encounter one of these on fruit trees, that will so resemble a short, stubby, dead twig, sometimes standing straight out, sometimes partially bent like an elbow, as to deceive the sharp eye of even the wren itself.

This instinct of insects, and especially of the beetles, to escape their bird enemies, led the late David Thomas, of Western New York, one of the best of the early horticulturists of our country, to use this canvas trap; and of all the many plans that have been employed it is the only one that has stood the test of experience. If the Curculio is to be conquered, the destruction of the embryo in the punctured fruit must be the chief remedy, and the canvas the adjunct.

This work would not be complete without a more circumstantial account of the Thomas mode of fighting the *Curculio*. The following, written by himself, is taken from the *Cultivator* of August, 1851:—

"It is more than twenty years since I caught this troublesome insect on sheets, and secured my crops of plums, nectarines, and apricots; and whenever the business has been thoroughly done, I have never been disappointed.

"An average of 1,500 Curculios, caught in the first ten days of summer, though sometimes rather earlier, have proved a sufficient reduction of the tribe.

"This method of protecting stone fruit I first published in the *New York Farmer*, and afterwards I several times introduced the subject into the old *Genesee Farmer*. Of late, however, I have seen reports of its inefficiency, and as the word 'shaking' has been generally used, perhaps the following extract from the latter journal, which I wrote in 1832 (vol. ii., pp. 155-6), may throw some light on the difficulty.

"The first statement was dated 6th Mo. 7, 1832, and describes the imperfect mode as commonly practised:—

"'On the first day of this month I observed some Curculios on the Plum trees in my fruit garden; and not knowing how numerous they might prove, or how much danger was to be apprehended from them, we spread the sheets which we keep exclusively for this purpose, and by shaking we caught from about fifty trees more than thirty of these insects. Since that time, on different days, we have made similar trials, but we soon became satisfied that only a few were left; and unless others migrate hither, which the *movement of the hogs will be likely to prevent*, I think their depredations will be very limited this season.'

"Three days afterwards I furnished the following statement, containing a very important improvement on the mode before described:—

"'Not three days ago I saw that many of the plums were punctured, and began to suspect that *shaking* the tree was not sufficient. Under a tree in a remote part of the fruit garden, having spread the sheets, I made the following experiments:—On *shaking* it well, I caught *five* Curculios; on *jarring* with my hand I caught *twelve* more; and on *striking* the tree with a stone, *eight* more dropped on the sheets. I was now convinced that I had been in error, and calling in the necessary assistance, and using a *hammer to jar the tree violently*, we caught in less than one hour more than 260 of these insects.'

"Now I should think that these statements would explain all the failures that have occurred in this business.

"At that time my trees were not large, but they have long since become so; and to attempt to shake them now, or to jar them with the hand, would be out of the question. *We strike them with an axe*, and the blows may be heard to a considerable distance. To muffle the pounder to prevent its bruising the bark, would be preposterous in the extreme; for the stroke, to be effectual, must be a sharp and sudden jar. A short stump of a sawed limb has been found best. Some of the success of these operations, however, depends upon the temperature of the weather. Thus, many of these insects fly off in the warm part of the day, and in the coolest mornings we catch them in the greatest numbers.—DAVID THOMAS, *Greatfield, 6th Mo.,* 1851."

After such clear and explicit instructions from the practical man, modified, improved, and perfected after twenty years of experience, it is unnecessary for me to say one word as to the *modus operandi* of the canvas remedy, except that probably the form recommended in the Plate will be an improvement on the plan of Friend Thomas, and the mop-stick for jarring the branches will be found greatly superior to the mallet or axe, where the trees are old and large.

As soon as the jarring is finished, go to work as you see the boy in this Plate. Get each Curculio between the thumb and finger, and kill it. Some have recommended to empty the whole contents of the canvas into hot water; this is not always effectual; a few will hold fast with their claws, some will fly away, or the water will soon cool. Better hunt them out individually, and crush them.

Early in the season many dried buds, the withered petals of the blossoms, and some insects, as Lady-bugs, will fall upon the sheet, requiring it to be turned or emptied for every tree. This can be done in taking up the sheet. The usual way of carrying this small canvas from tree to tree is to hold all the stretchers, long and short, in one hand, with the fold of the sheet hanging down. As you approach a tree, drop the long stretcher, and pass one of the short ones on each side, till the centre of the slip comes up snug round the body; then jar, and then crush, and so on, to every tree; and then begin again at the beginning, *and go on over the entire orchard just as often as you find Curculios.*

In cold, windy, or wet weather, the operations of insects are in a great measure suspended, and on such days the Curculio will require little attention. But as all insect life is active in proportion to the heat of the weather, when the hot days do come, and especially when the sun breaks out suddenly between showers, it will be found necessary to hurry this work.

The more vigorously the war is waged early in the season, the sooner it will be over. Watch the young fruits carefully as they approach the sizes of those on Plate II.; and the day that one of those crescent marks is seen, is the time to begin. Each female now will contain from twenty to twenty-five eggs, and she will want a fruit for each. Later in the season she will often be found with but few, sometimes only one or two, and, of course, she is not then capable of doing much more mischief.

If your trees are full grown, a larger canvas—ten or twelve feet square—will be required. This can easily be managed by one person, with the help of a small boy. Middle-sized trees can be jarred sufficiently with a common mallet, provided you can afford to cut off a good-sized branch, as shown at Figure 6 in this Plate. The edges of this stump should be carefully pared, so as to leave a convex surface to receive the blows. With proper care, such a stub will last during the season. The branch should not be less than an inch in diameter, or the stub would soon be split to pieces.

Old trees lose their elasticity, and cannot always be jarred enough with the mallet to cause the Curculio to let go; in this case the common mop-stick, used against the limb as shown in the Plate, answers perfectly. It should be properly padded to avoid bruising the bark.

The illustration on this Plate fairly represents the form of canvas I have always used. It is made of common strong *white* sheeting, and if properly taken care of will last for years. Many other forms have been described. Some use a common sheet, and that will answer for a very few small trees in a garden; larger ones would require two; but the stretchers will save much time, and will be found indispensable if the weather should be windy. One gentleman, whom I know, has arranged his canvas upon a frame, fitting loosely so as to sag to the centre, that the Curculios may roll down together; and this frame is fitted to a two-wheeled barrow, the slip projecting beyond the barrow, and that part of the frame thus brought against the body of the tree made strong and padded for the purpose of giving the necessary jar—a butting machine—something in the goat style. Such a contrivance, if made strong enough, will save the labor of one hand.

Most people of our country, who have attended the large agricultural and horticultural exhibitions, have seen the contributions of plums sent by Ellwanger and Barry of Rochester, N. Y. Often twenty, thirty, or even more varieties will be found on the tables from their orchards. Pears, apples, and grapes are contributed by many others, but they are often the only exhibitors of plums.

The *Horticulturist* of 1859, page 527, contains the following:

"*Curculio Remedy.*—The *Valley Farmer* publishes the manner which Ellwanger and Barry, of Rochester, take to rid their fruit-trees of this enemy. They employ two men, whose regular business it is to carry out this operation. A light wooden frame is made, on which canvas or cheap muslin is stretched, made large enough to cover the space under the branches of one half the tree. Also a similar one to occupy the remaining space. A branch of the tree has been previously sawed off, thus leaving a stump three or four inches long (one inch, only, would be better). After the Curculio catchers are placed beneath the branches, which can be quickly done, one of the men with the mallet strikes the stump a sharp, quick blow. The little Turks drop, and are immediately removed from the 'catchers,' and the men proceed to the next tree. Many hundred trees can thus be gone over in a few hours."

Twice last season I visited the plum orchard of these gentlemen. I certainly saw nothing in Western New York so beautiful, nor have I tasted anything so good as their plums. They have all the best kinds, both old and new, two, three, or more trees of each. These are trained as dwarfs and planted quite close together. The use of the canvas in such an orchard is much more difficult than where trees are planted in the ordinary way; still it is used successfully, for every year they have plums. The Curculio is probably as common here as in other places. In a careful examination of the pears and apples on their grounds, the unmistakable mark was as often met with as in other fruit establishments in Western New York. Charles Downing resorts to similar means of saving his plums.

I know a gentleman in New Jersey who not only has the taste for fruit-growing, but the energy necessary for conquering the enemies of fruits. You see in his grounds, in the most perfect order, pears, apples, grapes, cherries, and all the smaller fruits; but he takes no special interest in pointing them out to visitors. They are common—others have them; but when you go with him among plum, apricot, and nectarine trees loaded with fruit, you see his consciousness of triumph, and oh, how beautiful they are! Why will not every one who has retired to the country for enjoyment have such enjoyment as this, by conquering these enemies, and making for himself a primeval Eden, with all the modern improvements? Mother Eve managed the fruit business badly in her day, and gave Adam, and all the rest of us, a great deal of trouble; but if we face that trouble resolutely we shall find a recompense.

Lists of Plums to Choose from.

I.

From Charles Downing.

Bradshaw,	Bleecker's Gage,
Coe's Golden Drop,	Green Gage,
Denniston's Superb,	Imperial Gage,
Jefferson,	Lawrence's Favorite,
Lombard,	McLaughlin,
Prune d'Agen,	Purple Favorite,
Royal Hative,	Reine Claude de Bevay,
Imperial Ottoman,	Jaune Hative,

Washington,
Magnum Bonum (Yellow),
Red Gage,
Schenectady Catherine.

Yellow Gage (Prince),
Parsonage,
Schuyler Gage.

To which I would add Pond's Seedling, Victoria, Fellenberg Prune, Pear Prune, and the Mellen Gage. The last a remarkable plum, both as to quality and for its long period of ripening.

Apricots.

Breda,
Peach,
St. Ambroise.

Large Early,
Moorpark,
Purple or Black.

Nectarines.

Boston,
Early Violet,

Downton,
Elruge.

To which I will add Early Newington and Stanwick.

II.

American Pomological Society's Catalogue of 1864.

Bleecker's Gage,
Coe's Golden Drop,
Damson,
Early Favorite (Rivers),
German Prune,
Green Gage,
Hulings' Superb,
Jefferson,
Lombard,
Monroe,
Peach Plum,
Purple Gage,
Reine Claude de Bevay,
Royal de Tours,
St. Martin's Quetsche,
Washington,
Yellow Gage (Prince's),

Bradshaw,
Columbia,
Duane's-Purple,
Fellenberg,
General Hand,
Goliah,
Imperial Gage,
Lawrence's Favorite,
McLaughlin,
Orleans, Smith's,
Prune d'Agen,
Purple Favorite,
Royale Hative,
St. Catherine,
Victoria,
White Magnum Bonum.

American Pomological Society's List of Apricots.

Breda,	Early Golden,
Large Early,	Large Red,
La Fayette,	Moorpark,
Orange,	Peach,
Red Masculine,	St. Ambroise.
Turkey,	

American Pomological Society's List of Nectarines.

Boston,	Downton,
Early Newington,	Early Violet,
Elruge,	Stanwick.

I have nothing to say on the cultivation of fruit trees in general. The instructions in the books are ample. But upon the management of the Apricot Orchard some account of my own experience may be appropriate here. I commenced with ten trees, five from a North River Nursery, and five from André Leroy, France. The former had been budded near the ground on Plum roots, and were Apricot trees. The latter were budded five or six feet up on Plum trees—making Apricot tops on Plum stocks. The former never grew large and soon died—the latter grew as large as Plum trees and were long-lived.

A few years' experience proved, that in that particular situation, as to climate, soil, and exposure, I could have as regular and full crops of apricots as of plums, and with no more trouble from the Curculio. I changed whole orchards of young plum trees, then just beginning to bear, by budding; inserting the apricot buds into all the branches of the young plum trees. The success was perfect; and in the third summer such trees were bearing very valuable crops—the fruit large and beautiful.

The trunks of some Apricot trees, like those of some Cherry trees, in our climate, suffer badly from extremes of weather. This may be guarded against by budding high up on the Plum. The Apricot bud early in the winter will bear a greater degree of cold than the Peach bud, the latter being killed at a temperature of 18° below zero. But the Apricot bud will begin to swell with the early warm days of spring before the Peach bud shows any change, and then the liability to injury from cold becomes reversed.

By blossoming so much earlier than other fruit trees, the Apricot is considered too uncertain to be depended upon as a crop in most parts of our country; but if situations were chosen, such as are generally selected by apricot growers in Persia, Italy, and France, it would be found to flourish just as well here as there. It does not require warm countries, neither should the climate be very cold. But always plant on your high hill-sides; on mountains, if you have them. And if in such a situation that the sun shall not strike the trees until an hour or more after rising, so much the better; frost on the blossoms would then have time to evaporate, and the heat of the sun's rays could do no injury.

In the cultivation of the Apricot as a market fruit, much of its value will depend upon the perfection of its ripening upon the tree. Like the Green Gage and other superior plums, it should remain upon its native-stem until the exact time has come—no pulling to take it off, only gently coaxing. During the last two or three days that such fruits hang upon the trees, before they become what is called dead ripe, they increase greatly in size, and the richness and beauty of their colors are then fully developed. Fruits thus perfectly ripened are always wholesome, and are as superior to those picked prematurely, as the blackberry which surrenders to a touch is to the one torn off when only red.

Some may suppose that such fully ripe apricots, plums, or nectarines, cannot be carried any distance to market. That will depend upon the packing. They must not be bruised, of course; and if that is guarded against it will be found that fruit fully ripened will keep better than that gathered prematurely. It will not wilt, and is not so liable to rot. Small baskets are best. I had them made one foot long by six inches wide and four deep. Two, side by side, made a square. Two on the top of these, crossing them, and then two more—the six going together brick-fashion—made a very convenient package for a small-sized square crate. Two rows of the Moorpark Apricots, packed with intermediate layers of their own leaves, would nearly fill one of these baskets level with the top, and fifty apricots were the usual allowance. The six baskets contained about half a bushel. The crates were made entirely of slats, the lids fitted with hinges and locks. If carefully managed they will last for years. Some crates were made double this size, to contain two packages of six baskets each.

Few sights in the markets are more refreshing than the opening of such crates filled with perfectly ripe Moorpark Apricots, picked the day before. The extra expense caused by the Curculio is soon paid for when you can sell at such rates that a bushel comes to forty, fifty, or sixty dollars, as it would now.

One of my young Apricot orchards was an object of special interest on several accounts. It was so situated that all the Curculios that attacked the young fruit had to come from a distance. Every day some of these would be found on the outside rows, but so systematic was the warfare made upon them, that they never got within these rows. Every punctured fruit in that orchard was destroyed, but the next year it would be the same thing, the Curculios coming from a distance. The boys with the canvas would go over that orchard, with others, in the mornings, and I would make my calls at intervals during the day, to assure myself that no mischief was going on. Occasionally a Curculio could be seen at work, and then I would experiment with the jarring process, beginning with a gentle shake, then a harder shake, then a very gentle jar such as a blue-bird or bob-o'-link would make alighting on the tree in its cautious fluttering manner, then the decided jar of the robin or oriole. Such experiments led me to the opinion that the Curculio has an instinctive fear of birds.

I have often watched this insect on the ground after the taps had caused her to let go. She will lie on the bare earth or among the grass as quiet as if dead, as long as the danger lasts, and then unfold her legs, that had been closely drawn up, and creep off towards the tree. Soon you will see her moving rapidly up the body, on the limbs, and out on a branch of that tree. If the weather is very hot, the speed of her motions will be extremely rapid. If any one who reads this book is at a loss for employment, let him plant some high hill with an Apricot orchard, protected from the east, and keep off the Curculio. It may be made both pleasant and profitable.

Nothing has been said as to the jarring process for saving Apples, Pears, or Cherries. Young trees of these fruits just beginning to bear may be easily jarred with the hand; but larger trees will be found very unyielding to anything except the mop-stick to the branches. The Plums, Apricots, and Nectarines will be enough for any one to take care of by this troublesome process; and if all the young fruits on the farm, and especially on all the neighboring farms, have been properly disposed of the year before, there will be little occasion for its repetition. What the Curculio will then take of such fruits will hardly be missed. I have sometimes found great benefit from jarring young peach orchards for three or four days, and especially the trees bearing the very earliest kinds.

PLATE VIII

PLATE VIII.

1. A twig from a Plum Tree that had been profusely washed, or syringed, with what has been called the "Whale-Oil Soap Mixture."
2.2.2. Three Plums on this twig, painted exactly as they appeared June 24th, 1863, each Plum showing that it had received a portion of the mixture in the general syringing that the tree had been subjected to.
3. A Curculio as seen at work at the time.

PEOPLE who have had the fewest opportunities of acquiring information on the subject of medicine, and who are often totally ignorant of the science, are those most likely to be the dupes of quacks; and the vast sums received by the publishers of newspapers for inserting quack advertisements, testify to the number of such victims.

On the subject of Insects nearly all, educated and uneducated, are alike ignorant. Few have the time, still fewer the inclination, to devote much thought to such a subject; and those who do, usually study it strictly as a science. They arrange insects in orders, classes, families, genera, and species. They learn their Latin names, and what those names signify—in fact, become entomologists. Such people are often enthusiastic collectors of insects, and much of their time is pleasantly employed in arranging them in cabinets. The discovery of an undescribed species is as gratifying to an entomologist, as the finding of a new plant to a botanist, or a new fish to Agassiz. But these investigators seldom inquire which are the useful or which the injurious.

Others become interested in studying the habits or instincts of particular classes. The elder Huber spent forty years in the study of Bees and Ants alone, and left the subject unfinished. But his works have made many friends for these two species, who are continuing the investigation, and the general interest in them is constantly increasing. The great work of Kirby and Spence has made a friend to insect investigations of every one who has read it.

In our country, Harris and Fitch have laid broad foundations for usefulness—

too broad, perhaps, to complete the superstructure, but greatly useful as far as they go. Their works have created an interest in such studies that is doing much good. But still a vast majority of people—even religious people—have such a prejudice, such a repugnance to insects, that they do not hesitate to crush all indiscriminately. Such people seize with avidity all the nostrums they see recommended, especially in the Agricultural papers.

As the science of Surgery emerged from the deep darkness of the early ages, an eminent physician wrote, "Millions have died of medicable wounds." It might now be written that hundreds of millions have died from nostrums prepared by men who knew little of medicines, and still less of the human system.

Of all our insect enemies none have had so many remedies proposed for their extermination as the Curculio. For twenty years I have been making collections of these, and I cannot imagine anything of less value, unless it should be a similar collection of quack remedies for consumption or rheumatism. A few of these I propose to introduce here—some to be examined seriously, and some to be laughed at. Many of the newspapers of the large cities publish country editions, some of which have an immense circulation. To make these more useful to farmers, a column or two is often devoted to agricultural reading. Many of the religious papers have a similar department. One of these latter, published in New York, has printed annually for several years a Curculio remedy, of which the following is a copy:

"We have received requests from several persons to republish the directions for preparing and applying this mixture. The following are the proportions. If any are unable to obtain the whale-oil soap, strong soft soap may be used.

"THE MIXTURE.—To one pound of whale-oil soap add four ounces of sulphur. Mix thoroughly, and dissolve in twelve gallons of water.

"Take one half peck of quick lime, and when well slacked, add four gallons of water, and stir well together. When settled and clear, pour off the transparent part and add to the soap and sulphur mixture.

"To this mixture, add four gallons of strong tobacco water. Apply this compound when thus incorporated with a garden syringe to your plum or other fruit trees, so as to drench all parts of the foliage. If no rains succeed for three weeks, one application will be sufficient. If washed by rains, it should be renewed."

I find attached to the above quite a number of articles clipped from Agricultural papers, alluding to this remedy; and among them the following from the *Country Gentleman*. It was an extract from a letter from Maryland:

"As to the Curculio, I am dead beat—but not subdued. A remedy vaunted in a New York paper had my confidence for two years, in each of which the frost killed my plum blossoms. In the third the plums appeared, and so did the inevitable Curculio. Nothing loth to encounter him, I mixed my nasty ammunition—whale oil, soap, tobacco, sulphur and lime—seized my squirt, and charged the enemy in front, flank, and rear, windward and leeward, right, left, and perpendicularly. The consequence to the Curculio did not seem important—perhaps he rather enjoyed the aspersion—but I got not a single plum."

In the report of the Yale Agricultural Lectures, Dr. Fitch alludes to this mixture, and thinks that two of the ingredients may be useful.

To ascertain positively whether this mixture had any effect in repelling the Curculio I submitted it to many tests. The Apple, Fig. 9, Plate II., was the subject of one experiment; but as that was punctured when the insects were in confinement it was not conclusive. I next washed a branch of a Plum tree profusely with this mixture, prepared exactly as recommended, and liberated a number of Curculios upon it. They commenced the puncturing operations on the plums coated with the fluid with the same avidity as upon the unwashed ones.

Next, I watched some Plum trees in the garden of a neighbor who used this mixture most perseveringly through the season, and they were just as much punctured as were the plums in neighboring gardens, where nothing was done to protect them. To settle this point still more positively, I made a visit to the Agricultural editor of this paper, at his country place in Westchester county, New York. The mixture was prepared here by the barrel, and used profusely not only on the Plum trees but on rose bushes. The gentleman being absent at the time, while waiting his return, I experimented with the gardener on various insects. After the mixture was thoroughly agitated, so as to be almost thick with the ingredients, Curculios were put into it and kept under, and then, when they would come to the surface for breath, were forced under again and again. But still these Curculios, as soon as they had time to rid themselves of some of the mixture, would creep away, and when fairly dry would unfold their wings and fly off.

Several slugs from the rose bushes were treated in the same way, and although they seemed to be more inconvenienced than the Curculios, they all survived it. The rose bushes at this time, June 24th, showed the presence of very few of these slugs, and it might have been inferred that the mixture that had been used so freely had killed them; but at this time of year this pest has generally come to maturity as a feeding insect, and has formed its cocoon, preparatory to its transformation. The leaves on the rose-bushes in this garden were as much injured by this slug as in other gardens where nothing had been used to prevent their ravages. Rose slugs, like many other insects, appear with considerable regularity, and nearly all disappear about the same time. Remedies used at the latter period will apparently be effectual, and thus acquire a reputation to which they are not entitled.

Soon after these little experiments were concluded, the gentleman arrived. We made a tour of observation, during which I was able to point out the Curculio of which Fig. 3 on this Plate is a representation, in the very act of mischief, in spite of the mixture with which the tree had been deluged.

In a pleasant ramble through these beautiful grounds I discovered one great source of the number of Curculios here. The roads and avenues on the lawn were lined with shade and cherry trees interspersed. These trees were now old and large, and the cherry trees had a fair crop of fruit. Birds were present in great numbers, but not enough to eat so many cherries; probably not enough to eat that portion of them containing the embryo Curculios so often found in this fruit.

This gentleman, who has had such undoubting faith in the efficacy of this mixture as a remedy to protect fruit from the attacks of the Curculio, has probably had plums some seasons after using it, as many others have had after applying the various other remedies that have been in vogue. Often when they have such crops, had they examined their neighbors' trees where nothing had been used, they would have seen just as many plums. I have shown why we occasionally have such seasons of abundance. The general or partial droughts of the preceding season killing the embryo Curculio during its transformation, is one of the contingencies in the life of this insect, that has probably given to many of the Curculio remedies their adventitious reputation.

Much has been written about planting fruit trees so as to lean over water as a way of preventing the depredations of the Curculio. On the 25th of July, 1863, I was one of a party to visit the vineyards of Dr. Underhill, at Croton Point, on the Hudson River. That gentleman had solicited the appointment of a committee at a meeting of fruit-growers, to examine his mode of cultivating grapes. The visit was a most pleasant one. The number of grapes, and the manner of cultivation, were subjects of general admiration. The diseases of the vine and fruit, and some of the insect enemies, will be spoken of hereafter.

While here, we visited the Doctor's Plum trees planted round an artificial pond. They stand at an angle of about 45°, and so close to the edge of the bank that the greater part of the branches are over the water, so that when the fruit comes to maturity on these trees a boat will be necessary to gather the greater part of it. In a very careful examination of those trees having fruit on at this time, we found it badly punctured by the Curculio. On the plums high up on the trees, and especially on those branches leaning furthest over the water, it was impossible to see whether the crescent mark was there or not; but wherever near enough to be examined, we could see no difference between those plums hanging over the water and those over the land. They were just as badly marked with the punctures of the Curculio as were the plums on some trees at the neighboring station of Croton; just as badly stung as in Newark and other places I have visited this year on purpose to see the extent of the ravages of the Curculio. Gentlemen who have often seen these trees other years, have told me that they have always had a similar experience.

Dr. Underhill, like others, has had crops of plums, and these crops have probably been ascribed to the circumstance that they grew over water; and he believes that the merit of the plan is attributable to the sagacity or instinct of the insect: *That she must not deposit her eggs in fruit so situated that it will fall into water.* To carry out this theory, it would be necessary for the Curculio to know that the plums in which she deposits her eggs *will fall* from that tree; that if they fall into the water, the grubs they contain *will perish;* that if they fall on land they *will be safe.* The question here arises—Has the Curculio such instincts, or such sagacity?

In this world of wonders in which we live, there is nothing so wonderful as the instincts of insects. The impulses that control their actions are strangely perfect. They are no more likely to go wrong than a machine. We do not know what instinct is. We cannot define it. No matter how we put words together, they will give no adequate idea of what this blind impulse is. We cannot weigh, measure, see, or feel what is called gravity. But it is that *something* that keeps the universe in order; that *something*, in the ordering of the *Almighty*, that prevents one world from jostling another, and creation from falling into confusion.

Who can understand how the *Cicada septendecim*, after passing nearly seventeen years underground, should come to the surface in the evening of a certain day of the month, with almost exact regularity, generation after generation, for centuries? How should a certain kind of wasp know, that when she builds a cell of mud for the reception of her egg, she must put in a supply of insects for food for the young that will be born of that egg, and that on a certain future day she must break open that cell, and give her young a fresh supply? Who teaches the neuter bee—that nondescript that cannot be a parent—how to fabricate a cell for the young of another? Such curious instances of the instincts of insects could be multiplied till they would fill a volume, and all would be wonderful—equally beyond our understanding, but all consistent with their wants, and in accord with the rest of nature. Those who carefully observe these things will feel that they are in a world overruled by an *Omnipresent Guide* of all things. But the *Superintending Guide* that teaches the little Curculio to deposit her eggs in fruits where the future young will find food, would hardly give her an instinct to guard her against depositing that egg where fruits never grow except on trees planted contrary to nature.

We were told to-day that the tides were sometimes so low as partially to drain this pond, and it was then the Curculio punctured the fruit over where the water should be. The same special instinct that would teach her to avoid the water, should also admonish her to avoid the danger of the tide-water mud, the one being as fatal to the future grub as the other.

Planting fruit trees in this way will certainly diminish the number of Curculios; but as long as millions of young apples are permitted to lie undisturbed on the ground in the orchards in the neighborhood, to bring forth their vast armies for the next year, it will hardly be worth while to dig such ponds and plant trees round them in such an awkward position for the little good they would do. The embryo Curculio in the fruit that falls into the water will perish undoubtedly; but that water, or the fear of it, will not prevent the parent using that fruit. The teachings of instinct are so exact and unvarying that one punctured plum over water explodes the theory; and if the theory is correct, a tub of water under a tree must protect a column of plums of the tub's circumference from the bottom to the top of that tree, and that certainly would be a curiosity with some of the light-colored, full-bearing varieties.

It is not at all likely that many will plant trees in this way; but as some have done so, I have been thus explicit on this point, to guard others against such an expensive and awkward way of trying to outgeneral the Curculio, since reason and observation teach us that it is of very little value.

Salt and Lime as Curculio Remedies.—In Hovey's Magazine for 1851, C. Goodrich, of Burlington, Vt., gives the following experiments: "Flower-pots were filled with garden soil, on which a layer of fine salt, a quarter of an inch thick, was deposited. On this bed of salt were laid punctured plums containing grubs of the Curculio. The grubs came out of the plums, and passed down through the salt into the soil, from which perfect Curculios emerged some weeks afterwards. The same result took place when fresh air-slacked lime was substituted for salt, and where soil alone was used. The pots being exposed to the weather, the salt was soon washed into the soil, but there was no difference in the appearance of all the insects."

I have often performed such experiments, using the lime and salt not only singly, but mixed. Ashes have been tried, flour of sulphur, snuff; but none of these seemed to interfere with the safe transformation of the grub into the perfect Curculio. The placing of such flower-pots in buildings where the earth would become perfectly dry, gives a serious check to this process. My experience has been that almost every one perishes, and this circumstance should be constantly borne in mind when we are investigating the Curculio remedies, about which the testimony is so conflicting. There are seasons when the ground becomes so dry that almost the entire generation of this insect will perish from this cause. The next year the plums will be but little injured wherever the drought prevailed the year before. Remedies used in such seasons may possibly receive the credit they are not entitled to from many who have not the knowledge to trace an effect to its cause.

About the time that I was commencing the great battle with this insect for the protection of my orchards of Plums and Apricots, I searched the books and agricultural papers most carefully for Curculio remedies. Washes containing lime as the chief ingredient were often found. In several numbers of the *Horticulturist* there were communications from T. W. Ludlow, Jr., of Yonkers, on the Hudson River, N. Y., so circumstantial that it really seemed as if the long-looked-for remedy had been found. I tried it faithfully. A coating of this lime mixture, thick enough to make a plum look as a man does when his head is being moulded for his bust, would protect them, but nothing short of that would. Ordinary whitewash was not regarded in the least. The next year, about the middle of June, I visited Mr. L. to ascertain how it was that he and I should have such a different experience. His trees were white with this lime mixture—they had been fairly deluged with it; but the fruit was badly punctured. Fresh marks were everywhere visible, gum was exuding from others, and many were already on the ground containing the grubs well advanced in growth. Mr. Ludlow acknowledged promptly that it had proved a failure that season, and seemed very much at a loss to understand it. At that time I was not aware of the influence of droughts, and did not inquire whether the season, when his remedy had apparently been successful, had not been preceded by such weather.

Whether Mr. L. ever published an account of this failure I do not know; but he certainly should have done so, that those who had been induced to try the mixture from his recommendation should have been undeceived. A prompt report of failures is often of more value than the reports of success.

In the *Cultivator* of May, 1851, will be found the following on the subject of *Lime for the Curculio*, by the Horticultural Editor, J. J. Thomas:

> "Much having been said in favor of lime as a remedy for the Curculio, and as the time is approaching for its yearly assault on young fruit, the knowledge of past experiments becomes desirable. A near neighbor —who is a distinguished fruit raiser—tried lime in nearly all imaginable ways last year, and with the following results: Nectarines, Plums, and Apricots were thoroughly syringed with thin lime wash; and as each successive rain and heavy dew carried it off from the smooth surface of the young fruit, it was re-applied as often as necessary. Special attention was given to the Nectarines, which for six years of blossoming had yielded no crop; and to be still more secure against this, the lime was applied carefully with brush to each young Nectarine. About three days in the aggregate were spent in this way; and the result was, that the full number of six entire specimens of the Nectarine were saved from destruction out of the whole orchard. But on further inquiry it appeared that these six all grew on a tree under which a young calf was kept confined during the season of operation; and to whose presence, chiefly, these specimens owed their escape.
>
> "The lime was believed to have a repelling influence, and some hopes were at first entertained of its efficacy; but it was soon discovered that the coating was disregarded, and the eggs were thrust through it into the green pulp. The whole trees, with their entire crop of leaves whitened with lime, did not present a very ornamental appearance.
>
> "The application of lime appears to have been elsewhere in some cases quite successful. It becomes a subject for inquiry whether any collateral influence assisted it; whether the favorable result was not owing to something else, and was erroneously ascribed to the lime."

Friend Thomas, with such evidence before him, should have taken the responsibility of pronouncing judgment in this case, and announced that lime was of no use

Paving to prevent the Curculio.—This is one of the remedies about which so much has been written, and some who have tried it testify so positively as to its usefulness, that it will be proper to devote some space to its consideration.

In the *Horticulturist*, Vol. iv., p. 62, 1849, Lewis F. Allen, of Buffalo, New York, gives a long article about Lyman A. Spaulding's plan at Lockport, which he considered successful. The Editor of the *Horticulturist* (Downing) testifies to the plan as being excellent.

At page 128, of the same volume, H. W. S. Cleveland, of Burlington, New Jersey, writes to show that "*Paving* is not a *preventive*." And the Editor says it will not answer in *all* cases. Probably only where the Curculio is not very abundant.

The following is from the same volume, p. 244 :

"*Paving Plum Trees.*—Mr. Downing: Paving about Plum trees to thwart the Curculio always appeared to me to be the sheerest nonsense. Whether they fly or not, the value is precisely the same. It presents no obstacle to wings, and certainly facilitates progress to legs. Why won't gentlemen who are so fond of recommending it, try the same experiment with their Cherry trees, to keep off the birds, or build bridges over their garden fences to prevent the inroads of unruly boys, and enrich us with the record of their sagacity? If the Curculio passes up the body of the tree, as is claimed to be 'conceded,' why not invest funds to the extent of a cent per tree in tar? A sum not so exorbitant but most plum-growers might be tempted to risk it; and the expedient would certainly be more embarrassing to footsteps than brick pavement.

"The true worth of a recommendation for the preservation of fruit, in addition to being effectual, is its capability of universal application. Paving, at $3 per tree, is as generally impracticable, and would be as rarely adopted, as inclosing trees in glass houses. Its expensiveness and doubtful utility at least condemn it.

"There is nothing known of the nature of the Curculio opposed to the probability that, having wings, they fly, if necessary, to perpetuate their race; nor of the grub, that, having legs, they also have wit enough to convey themselves, after escaping from the fallen fruit, from where they can't burrow to where they can, though in passing over pavements they sometimes doubtless perish, like bigger worms over the deserts of Sahara. The sagacity of the Curculio provides against this difficulty, where it is possible, by the selection of localities more favorable to the prosperity of the rising generation.

"*Hence, paved trees are shunned where others can be found.* Let all be paved, and all would suffer. I venture to predict that, unless Mr. Spaulding of Lockport has near neighbors, whose plum trees are not paved, every one of his own will hereafter be attacked. By cultivating two sets—one for himself and another for the Curculio—he has hitherto preserved his share; but by paving the whole, I am mistaken if he will not be the loser. Paving, on such terms, may by some be considered advisable; but I don't apprehend a scarcity of brick will grow out of it.—J. C. H., *Syracuse, Sept.* 18, 1849."

At page 246 of the same volume, Mr. Longworth, of Cincinnati, commences a defence of the paving plan:

"*The Curculio.*—A correspondent of yours, for a single year, tried paving to save his plums from the Curculio, and failed; and therefore concludes, paving is not a preventive. He is confirmed in this opinion, because 'the insect has wings; and presumes, as the pavement insured a crop with Mr. Allen, that his plums belonged to the Dutch family.'

"It appears to me singular, that persons will, from a single year's experience, undertake to express an opinion. I have for twenty-two years had about twenty plum trees surrounded by a brick pavement, and have never failed to have a crop of fruit from them. A few of the fruits, in some varieties, are occasionally stung by the Curculio. In my adjoining grounds I have as many trees of the same varieties; and two years out of the twenty-five, have had a fair crop of fruit. The other twenty-three years the Curculio left not a single plum. The safety of the fruit in a pavement does not arise from no Curculio being bred in the ground. If a person does not raise them his neighbors will give him a liberal supply. As an experiment, I planted a small plum tree, 1000 feet from any plum tree. The first year of its bearing every plum was stung by the Curculio, and for years after. The safety of a pavement arises from the instinct of the insect. It will rarely deposit its eggs over a pavement; as the young, when they fall from the tree, cannot secure winter quarters in the earth. The mother feels too strong an interest in the children to subject them to such a fate.—N. LONGWORTH, *Cincinnati, Ohio, September,* 1849."

In the *Horticulturist*, vol. vi., page 243, N. Longworth speaks again of the paving plan—thinks the Curculio is timid, and afraid of pigs, poultry, and people. And in the same volume, page 374, Mr. L. alludes again to his practice of paving; repeats that the insect is a timid one, and says the proximity of his trees to the house, where persons are constantly passing, *may aid* in keeping off the Curculio.

In the same volume, page 383, Wm. Quant, gardener to W. C. Langly, Esq., Third Avenue, Long Island, has a short article. He says he has had a long battle with the pest, and when he sees the accounts of success, he wants to be invited "to come and see and believe." He says also that he was at one time a gardener for Mr. Longworth, and that the reported success of the pavement plan there was not true.

In vol. viii., page 428, the Hon. James Matthews comes out with a remedy, but it having cost so much time, etc., etc., he wants something before it can be made known. There are numerous allusions to the Matthews plan. The New York State Agricultural Society appoint a committee to examine it, and the editor of the *Country Gentleman* is one of the members. The people become impatient, and call upon the committee to report; but the *Country Gentleman* replies—that the plan is a secret and cannot be explained—that they want more time, but recommend the pigs and jarring process. I have been able to find a few more allusions to this Matthews remedy, but have never seen the report of the Committee. Once in the garden of a friend I noticed a peculiar kind of pavement under his Plum trees, apparently made of small stones, or chippings from a quarry, and cement. I was left to infer that it was a specimen of the Matthews plan—but it was a secret.

The whole system of paving as a remedy for saving fruit has so generally gone out of use, of course from its want of success, that it will not be worth while to say much more about it. My near neighbor, Mr. Pierson, tried it most faithfully under the immediate directions of Mr. Longworth himself (who often visited Newark), but without the slightest effect.

This, like planting over water, was supposed to be effectual, in consequence of the instinct of the Curculio teaching her not to deposit her eggs where her young would be in danger. We have seen that the Curculio throws herself to the ground when disturbed. If she falls upon a pavement, and poultry are about, she is certainly more liable to be caught. So also with the grub. It passes into the ground at once upon leaving the young fruit—if it can; but upon a pavement that would be difficult—it would be a longer time exposed to its enemies. The grub of the Curculio laboriously at work on a pavement in a poultry-yard, would have a poor chance for life.

But that the Curculio will abstain from the use of fruit because it grows over a pavement, I do not believe. It is certainly not an established fact.

In the *Cultivator* of June, 1852, is the following communication relative to the *Curculio in Michigan:*

"I propose to speak of the progress of the Curculio in Southern Michigan. I have been a resident of Lenawee county for the last eighteen years. The first depredations of this insect commenced about six years ago, the first season attacking a few only of our choicest Plums; the succeeding year they were more numerous, and since, continuing from year to year, puncturing every variety of Plums, and also Cherries, to considerable extent, and in some instances Peaches, and even Apples. All reputed remedies have utterly failed to save the fruit the last season. Previous to last year, those who were careful to jar their trees daily for two or three weeks, and to destroy the captured rebels, succeeded in saving a portion of their fruit. But the last season, this practice too was an entire failure, even when persevered in for months. In some sections of our country confining hogs in the plum orchards has been thought advantageous, and which has been the practice of the writer with signal success until within two years—my hogs being regularly fed under one tree, treading the ground so much as to destroy all vegetation—this tree retained its fruit until ripening, excepting last year.

"The cultivators of this fruit are entirely discouraged. One object in this communication is to inquire, through the *Cultivator*, if the Curculio has ever been known to absent itself from any district where it has been known to be prevalent—if not, then we may as well cut down our trees at once.

"Before the appearance of this insect, finer Plums were never grown, perhaps, than in this section, fine crops being obtained from grafting on the wild plum (*Prunus Americana*) in three or four years' time.

"ADRIAN, *February*, 1852. B. J. H."

The above is a remarkably straightforward account, and is valuable as showing the time of the appearance and the subsequent progress of this insect in a new country. Last year (1864) there was a section of country around Buffalo, N. Y., where only plums and peaches appeared to suffer from this insect, the apples escaping. I supposed the wonderful exemption of that neighborhood was owing to a partial drought in the season preceding; it was so like what I have seen and heard of so often before. I would let the trees stand, even if nothing else should be done to save the plums except to wait for favoring weather. It appears that the tree under which the hogs were so constantly fed was the last to yield to the enemy. David Thomas's account of his experience in jarring, as shown in the preceding chapter, will probably explain the cause of the failure when "persevered in for months."

In the same number of the *Cultivator* from which the above extract is taken, there is a short account from the *Prairie Farmer*, of a man who kept his hogs in his Plum Orchard for seven years in succession, during which time he had plenty of fruit, though none came to maturity outside of that orchard; but on changing the hogs to another part of the farm every plum was stung.

In the *Genesee Farmer* of 1848, p. 114, is an article from the *Horticulturist*, in which it is said that "a heap of fresh manure under the trees proved a remedy, and the Editor (A. J. Downing) gives his testimony to this plan, by stating that two Nectarine trees, standing by a fence near his stable, bore fruit to ripen, when other trees within a short distance shed all theirs in consequence of being stung by the Curculio." This question of the influence of manure or other strong smells under trees, for repelling the Curculio, has been more or less discussed from that time till now. I have tried the experiment often, but with no perceptible effect. It might here be asked, does any one know that a Curculio is conscious of a smell from a manure heap?

C. E. G., of Utica, writes to the *Cultivator*, March, 1850 :—" Having read, somewhere, that fresh stable manure put round fruit trees in flower would repel the Curculio, I put some round my Plum trees. As I had to take the manure when it was offered for sale, I was obliged to apply it a little earlier than I desired. Soon after a heavy rain fell, washing, of course, the soluble portion of the manure down upon the roots of the trees. Quite a number of valuable bearing trees died outright, and a number more were seriously injured. This was dear-bought experience." He says further, " I doubt the feasibility of this plan of repelling the Curculio. If the weather be dry or windy it can do very little good unless the quantity be large, and then you endanger your tree."

In the *Horticulturist*, Vol. x., p. 189, a friend of the editor proposes to flood the Peach Orchard once or twice a day, and drown the scamps; and the editor wants him to try it. In the same Vol., p. 357, Henry Croft, Vice-President of the Toronto Horticultural Society, C. W., recommends the use of the sulphuretted waters, such as that of Avon. These might be called the Hydropathic remedies.

In the same, p. 479, John Brush, of Brooklyn, N. Y., proposes branches of Tansy to be placed in the crotches of the trees. He had found it successful. But the Editor says in reply that all he had been able to save was by the use of Millinet. As to the latter, my neighbor, Mr. Pierson, has tried 100 yards of it at one time, enveloping the entire tree; but the Curculios would find their way in.

In the same, p. 431, Mr. J. R. Gardner, of Sunny Side, Montgomery Co., Va., piles small stones round his trees—it is successful. He does it because he has seen trees growing among stones in Pennsylvania.

In the *Ohio Cultivator*, 1850, vol. vi., p. 189, Z. Hampton, of Pennsville, Morgan Co., O., says: "Caleb Hall, a respectable citizen of Muskingum Co., thinks he has found a preventive for the ravages of the Curculio. Those wishing to save their plums, I think, will do well to try, and now is the time. His method is to melt brimstone, into which dip woollen rags cut into slips, say three or four inches wide and five or six inches long, stick them one at a time on the end of a pole of sufficient length, split a little at one end to receive them, set on fire about dark, and hold them burning under and among the bearing branches a few minutes, two or three evenings each week, for three or four weeks, by which he has saved his plums, so as to sell over sixty dollars' worth a year." I fear Friend Hall is one who is in the practice of killing his bees by fumigating with burning brimstone, and believes the Curculio can be killed in the same way. So it can. If he will make a box large enough to cover each tree, and as tight as a bee-hive, and then fill it with the fumes of his burning matches, until brimstone will no longer burn in it for want of oxygen, the Curculios will probably be in very much the same condition that bees are, after such a Satanic visitation; but the effect would be very bad upon the trees. There was some other cause besides an occasional smell of brimstone that kept the Curculios from Friend Hall's Plums.

The *Cultivator* of Sept., 1847, states, that—"Some time ago a remedy was proposed in the *Ohio Cultivator*, on the authority of Gen. J. T. Worthington, consisting of tubs, whitewashed inside, and containing an inch of water, placed under the trees in the night, with a lighted candle in each." The light attracts them, and it was averred that "*hundreds* had been caught in this way, in *one night*, in a *single tub*," and that it had been practised with much success by "one or more" fruit-growers of Chillicothe.

In a subsequent number of that paper, J. Dille, an intelligent nurseryman, states that he has "tried this remedy without any success whatever; that some of these insects were under water half an hour, without any apparent inconvenience; and that they ascended the side of the tub as readily as a sailor would a rope." Not many would have had the patience to try this experiment, as Mr. Dille did. We thank him for thus promptly proving and recording its worthlessness.

If there were a person connected with the Agricultural Department of the Government whose duty it should be to test the merits of new things, it might often do much good. I write a great deal by lamplight on summer evenings. I have caught hundreds and hundreds of insects that have been attracted by my light, but I never yet have caught a Curculio in that way. I have been in the habit, for years, of carrying a small vial of Curculios in my pocket. Sometimes I meet a person who talks as if he knew all about the Curculio. At the proper time my vial comes out for information, but the insects are seldom recognised. Even Agricultural editors do not always know what they are. Very likely General Worthington's friend would be equally at a loss.

On the same page of the *Cultivator* is the following short notice :—" A. J. Downing recommended, in the *Horticulturist*, throwing up the ground late in autumn in trenches and ridges, for the purpose of freezing them, and stated that a correspondent had found it quite successful. The writer tried this same way last fall, but this year they were thicker than ever. On one little tree of the Italian Damask Plum, not seven feet high, thus treated, eighteen Curculios were found at a single shaking." The grub of the Curculio goes into the ground several inches, and there it changes to a beetle, and this beetle comes to the surface, showing that as a beetle it has a power of making its way through the ground. Had Mr. Downing known positively whether the Curculio lives above or under ground in the winter, it would have been a beginning for an investigation of facts on which to found this treatment. If these insects live under ground in the winter—how far, exactly? and if brought to the surface in the autumn, would they not creep back again? or if they did not, would the winter kill them?

In the *Ohio Cultivator*, 1849, vol. v., page 42, George W. Dunn, of Chillicothe, writes:

"This is frequently called an age of improvements. It may be also called the age of oddities, one of which I will send you for the public benefit. I am acquainted with a Highland County farmer of the name of Martin, who is well known in the neighborhood for growing fine plums. A few weeks ago his son was at my house, and I asked him how they could raise such fine plums when no one else could. He replied, that as soon as the fruit was formed, they took a pocket-knife and made a slit through the bark, through the main stem and larger limbs of the tree, and this, he said, was all."

Now, strange as it appears, this was actually published in a respectable Agricultural paper, and the man who wrote it said he intended to try it himself. That was in Ohio, in 1849. Here is something from a Norristown, Penn., paper, in 1863 :

"*How to Prevent the Curculio from Destroying Plums.*—A perfectly reliable man who lives in this vicinity, was telling me, a few days since, how he managed to raise Plums. He says, just as the trees are coming into full bloom, he takes a ragged stone and bruises the bark in the crotches of the trees ; he leaves the stone there. That, he says, arrests the gum which will exude from the wounded place, and prevents its going to fruit, thus cutting off what he supposes to be the food for the larvæ. He says he has tried it for many years, and never fails when the trees blossom, except when he neglects to bruise. My informer says, do not be afraid of hurting plum-trees by bruising them ; he says the more they are bruised the more they will bear. Now, my friends, try some of your old lazy plum-trees; give them a regular trouncing, and report results."

We all know what Captain Cuttle said of his friend Jack Bunsby's judgments, "wisdom as is wisdom." Even so we may characterize a communication from A. C. Hubbard, in a late number of the *Michigan Farmer*, telling of some one who had been told by an "old Frenchman," that he must hang elder-bushes in his trees. He did it, and had plenty of plums. So Herman Dousterswivel, by means of "suffumigations," found a casket of gold and silver coins in Misticot's grave. To faith like this, the witch-hazel tells where to dig for water, and a horse-shoe nailed over the door insures good luck.

Those readers who have had the patience to follow me through the last few pages, may suppose that such articles can only be found in obscure newspapers. Read the following:

"*To Prevent Fruit from being Wormy.*—I have a communication to make in reference to the worm nuisance. You will, I think, receive the thanks of two cities by publishing the following:

"With a large gimblet or auger bore into the body of the tree, just below where the limbs start, in three places, a groove inclining downwards. With a small funnel pour a shilling's worth of quicksilver into each groove. Peg it up closely, and watch the result. Had it been done when the sap first started on its upward circuit it would have been more efficacious—yet, even now, it will greatly abate the nuisance.

"The plan was first tried for a wormy apple-tree by Samuel Jones, Esq., of Canaan, Columbia county, New York, and with entire success. It is believed that, far from damaging the trees, it will even add to the beauty of the foliage. In the case of the fruit above mentioned the cure was surprising, not only the fruit becoming perfect and beautiful, but the very leaf seemed to be larger and far more dark and glossy.

"Any one desiring further particulars (though no other is needed for doctoring our city trees), is referred to an eye-witness, the daughter of the above named gentleman, at 225 Union street, Brooklyn.

"Brooklyn, *May* 7. A Constant Reader."

When such nonsense can find its way into a paper like the *New York Evening Post*, it may be asked—What are the people to do?

Crude quicksilver is as harmless as water until it undergoes some chemical change, and this it cannot experience when plugged up in wood. But suppose it should change into corrosive sublimate, and so much of it get into the circulation of that tree as to poison it, how is it to interfere with the insects? They know enough not to eat poisonous food, and will not remain on a dying tree.

Every year many accounts go the rounds of the papers, that some one has saved his Plums by plugging sulphur in his trees. Others are equally successful by simply driving in nails. Here, too, faith is necessary—without it, they are certainly useless. Some cover the bodies of their trees with tar, that the Curculio may be stuck fast when travelling up and down. I have tried this, and a few have been caught, but tar soon becomes so glazed that they can travel over it without danger, and to be of any use it must be constantly repeated. Tar applied directly to the bark of a tree, even if only a narrow belt, will often injure, if not kill it. Cotton, or cotton bats, are sometimes bound round the trunks or larger branches of fruit trees, so as to make it hard travelling for these little insects. Some make little troughs of tin, and fill them with oil, and fit them to the trees, and all the Curculios that are so incautious as to fall into this oil will be killed. Insects have a breathing apparatus—air is as necessary to them as to us—it is the breath of life; the openings to their lungs are numerous and are on their sides—oil closes them instantly, and then they die. Had the Curculio no wings these last remedies would seem to be effectual in theory.

Plums are brought to the New York market in large quantities almost every year; some seasons in great abundance. The usual varieties are the Damsons, Horse Plums, Frost Gages, and other more common sorts. They are often packed in barrels, and although roughly handled they are generally but little bruised, being too hard for that. The plums alluded to in these pages would be as different from these as "Stump the World" peaches are from green persimmons. I am often asked where all these plums come from, and is there no Curculio there? They are shipped from Catskill and other places on both sides of the Hudson River, and are said to come from the mountains twenty and thirty miles back. There are neighborhoods in Albany, Schenectady, and Washington counties, N. Y., where it is rumored that plums always escape the Curculio. I have often visited places with such reputations, in the hope of finding the promised land, but have always seen more or less evidence of the presence of the Curculio. Still there is a difference in different localities, as to the extent of the injury done by these insects.

In the City of Hudson, N. Y., I have seen trees bearing fair crops of plums every year, even of the better and more delicate kinds. One old cultivator told me that the Curculio was rather useful than otherwise. It was there certainly, and took a portion of the crop; but always left enough, and those which remained were the larger for the thinning out. The soil here, as in many other places along the Hudson River, is the stiffest and most tenacious clay. Bricks are made there. Cisterns have been dug in this city that required no walls, the cement being applied directly to the clay itself. In this kind of ground the grub of the Curculio is unable to work far down, and such soils suffer more from droughts than any others. The Curculio, in passing through its transformation, will consequently more frequently perish in this than in other soils. In this way the comparative exemption of neighborhoods may be accounted for. We may often see the Damsons, Little Gages, Petite Mirabelle, and other varieties of the very small kinds of plums, that bear such prodigious crops, bring as large a portion to maturity as Cherry trees do, after losing much of the fruit from the Curculio; and persons not accustomed to investigate these matters, would be likely to come to the conclusion that these varieties were either entirely exempt, or that there was no Curculio there. If I intended to make a business of cultivating plums, I certainly would choose a clay soil, and the stiffer the better; and I should prefer that all my neighbors for many miles should also live on just such a soil, unless they would all unite with me in forming a Fruit-growers' society, that would thoroughly exterminate the insect pests that interfere with the success of this business.

In the *Genesee Farmer*, 1851, p. 96, we read that Mr. Harvey Green, of Jefferson Valley, Westchester Co., N. Y., "says he repels the Curculio by tying up straw in bundles as large as his arm, takes a long handle, sets the straw on fire, and passes quickly round the tree. They fly into the blaze and perish. Mr. Green says he gives it for what it is worth." This might have been worth something if Mr. Harvey Green had proved that he knew a Curculio when he saw it, and that he was sure that the insects that flew into the blaze and perished were genuine Curculios. My experience has been, that a disturbed Curculio does not often attempt to fly away, but falls to the ground.

W. N. Read, of Port Dalhousie, Canada West, writes in the *Genesee Farmer*, 1853, p. 125: "It would have done you good had you seen my Jeffersons, Washingtons, Hulings' Superbs, Green Gages, Columbias, Golden Drops, Apricots, and Nectarines, last year, all bending under a tremendous load of the finest fruits ever beheld in the neighborhood of Port Dalhousie, saved as follows: Placed two or three well-made wind-mills in the head of each tree, with a clapper attached to each, which struck upon a piece of sheet iron, and when the wind blew kept up a terrible jingling noise; one and a half yards of flag tied up so as to float nicely in the air as close to the tree as possible without touching it; and, lastly, when dinner was over each day, I would catch up a sheet made for the purpose, and say, 'Come, boys, hold the sheet,' and I would jar the trees, and kill all that fell upon it. Operations to commence as soon as the blossoms have fallen, and continued until the stone became hard in the fruit, after which the Curculio cannot make it drop, though some half or one-sided fruit will appear by his work, but will be small and hardly be missed." As wind-mills are at rest except when the wind blows, and as the Curculio does the chief part of her mischief in the stillest weather, I think wind-mills, even with clappers, cannot be relied on. And as to the flags, I do not believe that the little Turk cares one straw for all the flags that fly in all the British dominions. It was the "Come, boys," that saved the Jeffersons at Port Dalhousie.

In the *Genesee Farmer* of 1845, p. 91, is the following short article taken from the "*Maine Cultivator.*"

"The Curculio, or *Green Moth*, which commences its ravages on the Plum about the first week in June, by depositing its eggs in the fruit while it is yet in its infant state, can be easily exterminated by preparing a mixture in the proportion of a bushel of wood-ashes to a quart of soot and half a pound of sulphur, applied in the morning while the dew is on the fruit, in a sufficient quantity to coat the tree." And the Editor says, "The remedy presented is an easy one, and if effectual will be of great value. The Curculio has long and justly been considered one of the most troublesome depredators upon the fruit orchard, and its destruction is a consummation devoutly to be wished."

The above is replied to in the same volume, p. 103.

"*The Curculio.*—Mr. Editor: In your last number I saw an article copied from the Maine *Cultivator*, professing to give a 'remedy against the Curculio,' and names the *destructive* as a '*green moth.*' (!!)

"It is not a matter of wonder that every person does *not* know what a *Curculio* is; but it *is* a matter of wonder that constant readers of agricultural papers, most of which have again and again described and treated of this insect, and given engravings showing size, shape, &c., should not yet have 'made his acquaintance,' or at least have known whether he was a *worm* or *bug*. It is not a moth, or other worm, that does the mischief, as I have many times watched the Curculio and seen him perform the process, and this he does with the skill of a professor of surgery—first cutting a segment of a circle, and then depositing the egg, after which the juice exuding from the wound forms a 'sticking plaster.' I am very sceptical as to the exterminating properties of the *remedy* he gives (ashes, soot, and sulphur sprinkled on the tree), *it cannot reach the egg;* and as for the Curculio, he inhabits a sort of coat of mail, hard and resisting, and seems to care little for what surrounds him, *plums excepted.*

"The *remedy which will prove effectual*, if the gardener does his duty, is to *anticipate him*, and never let him exist; which is done if all the *punctured* plums that fall to the ground are burned, or given to the hogs.

"Now, sir, one word in relation to copying the article into the *Farmer*. I shall believe it was done without your supervision, as its erroneous description of the *Curculio* must at once have satisfied *you* that the writer knew nothing of the insect of which he wrote.   "Yours obediently,

"Rome, June 5.   J. H."

J. H. very properly reproves the editors of both these agricultural papers for publishing an article written by one, calling the Curculio a "green moth." Could all the foolish recommendations that find their way into the papers be as promptly met and as properly answered as this was by J. H., they would not do much harm.

This long account of Curculio remedies might be much extended. Many others have been proposed,- but, like the above, have been found wanting when fairly tried.

About ten or fifteen years ago there was great activity in the search for a Curculio remedy, chiefly with the idea of finding something available in connexion with the supposed instincts of the insect. We had the paving, planting over water, powerful smells, and heaps of manure; but all these, as well as the various mixtures for coating the young fruit, are now abandoned. The agricultural papers seldom speak of any of them, and few new ones are proposed. The destruction of the grub in the young fruit and the jarring process for killing the beetle during the season of mischief are all that have survived; and so little is now said of these, that most people have settled down into the belief that Nectarines, Apricots, and even Plums are to be given up. They say if these fruits could be had without trouble they would be very nice; but they can do without them. There are people who, if they find it troublesome to raise wheat, will live on rye or corn bread. But now, since the signs of the times indicate so plainly that even Apples must soon be given up also, unless we make fight against the insect enemies, perhaps the public will be aroused to a sense of danger.

I hope all who have followed me through this chapter on remedies, will resolutely determine that the question as to their usefulness is no longer an open one; that they at least are not to be depended upon; that the fight hereafter is to be directed to the killing of this insect, either as grub or beetle; that everything short of that may as well be given up first as last.

Here I would like to comment at some length upon many valuable articles which I have met with in the journals on the Curculio, but this chapter is already very much extended. Some remarks on the early history of the Curculio, by David Thomas, appeared in the August number of the *Cultivator*, of 1850. In this will be found a notice of a correspondence between Peter Collinson, of London, and John Bartram, of Philadelphia, about this insect, so long ago as 1736–7. D. Thomas also speaks of the early contributions of W. Bartram, Dr. Tilton, and the late Judge Darling, of Connecticut. The *Horticulturist*, especially while under the management of Mr. Barry and Mr. Mead, contained many valuable articles on this strange and important insect—one by the late Dr. Harris is full of interest; but really the very best, to my fancy, is from William Hopkins, of Pomona, Brunswick, Renssellaer county, New York.

EXTRACTS FROM A DIARY KEPT IN 1864.

As soon as I had determined upon the preparation of this book, I commenced a systematic investigation of the time of appearance, the habits, and the depredations of the various insects that would be likely to come under review, and have taken down, almost every evening, notes of what I observed during the day. During the season of 1864 this diary makes many volumes. That portion relating to each insect spoken of in this work will be introduced under its appropriate head. The part relative to the Curculio comes in here, and it will constitute a narrative of the important events of its career.

May 12.—Visited Trenton to-day. The Quince trees in blossom. The seasons are usually a week earlier here than at Newark, though there is not a difference of half a degree in latitude. Mr. Voorhees, President of the Agricultural Society, told me that they had already caught many Curculios by jarring the trees over a sheet.

May 13 —Caught three Curculios this evening by jarring a Green Gage tree. The Plums are now just forming; indeed many of the blossoms are not yet off the tree. Apricots are as large as the end of the little finger.

In a record kept for ten years in succession, near the Hudson River, latitude 42°, the time of the blossoming of the Apricot varied as much as three weeks—from the 11th of April to the 3d of May, but the young fruit attained the size at which the Curculio chooses to use it, on the 18th of May—not varying more than two days in all that time.

Many writers say that the war upon the Curculio must begin when the trees are in blossom. Had this advice been followed from the 11th of April to the 18th of May, on an Apricot orchard, it would have proved an almost total waste of time; probably few would have been found, as they do not concentrate till the fruit is of the proper size.

May 14.—The Curculios caught last evening are now exceedingly active. They appear to be of both sexes, and are as restless and full of life as birds are in the early days of summer. They had been placed in a wooden pill-box; and in holding it to the ear they can be heard moving about, both day and night. This evening I have been subjecting them to an examination under the microscope. The eye appears to have 147 lenses, and one of the females was found to contain twenty-five eggs. The head, twenty minutes after being separated from the body, was still in motion. The nippers at the end of the proboscis could be seen moving as if biting. A living insect being placed so as to show the under side of the body, gave a complete view of the ball and socket articulation of the head and legs, very perfect illustrations of that kind of joint. This accounts for the curved form of the mark made in the fruit. The body of the Curculio remains immovable while the incision is being cut, and, of course, it must conform in shape and size to the turning of the head on this ball and socket axis.

Portions of beetles are often used as settings for the microscope. The Diamond Beetle is very brilliant, but far less so than the Curculio. There are probably no combinations of colors so gorgeous as those exhibited in the wing-covers of a living Curculio under the microscope, where the rays of strong lights are concentrated by properly arranged reflectors. All the parts of this little beetle—eyes, limbs, and wing covers—develop bright metallic tints, while the minute hairs found on all parts of the Curculio appear as pearls. I have often wished that these vivid colors could be transferred to canvas; and my friend Hochstein has several times made the attempt, but he has now abandoned the undertaking as beyond his power.

One would think, in contemplating a Curculio, that it was as unpromising a subject to develop beauty as would be the head of a Toad to bring forth jewels; but "sweet are the uses of adversity," as Shakspeare says, in alluding to the contrast between good and evil, as exemplified in the Toad. In his days it was believed that there was a stone in the head of the Toad endued with singular virtues, and this was a compensation for the venomous effects of its touch. I am a firm believer in the doctrine of compensations, but the tongue is the Toad's jewel. All gardeners know the trouble we have with some insects, and few are more provoking than the striped bug on the young Melon vines. Place one of your pet Toads (and all gardeners should have such pets) among these vines, and watch him. See how like a streak of lightning that tongue flashes out, and how the Melon Bug flashes in. If you are a collector of insects, and want those that are active at night when you are asleep, take from the Toad early in the morning the supply he has gathered to ruminate upon during the day. Many of them will be as perfect as if you had caught them with your own gauze net. And what a variety! Yes, the Toad is ugly and venomous—to insects, but has a jewel in his head for us.

The Toad is a funny creature, and if you look at him as a philosopher should, without being angry, because he sometimes eats strawberries, you may find a great deal of amusement in him. I have seen animals go backwards into their burrows, but except the Toad, I have never seen any make their burrows backwards. Find one a little belated in the early morning—confront him, and watch sharply as if you suspected he had been eating strawberries or lady-bugs, and his eyes will begin to wink and blink; soon his head will be averted, as if he were ashamed, but all this time he will be settling away, going down as a canal boat does in a lock; his hind feet have been throwing out the earth from under him, reminding you of the description of the first steamboat—"a grist-mill afloat, with the water getting out from under."

Toads are fond of strawberry beds. The partial burrow beneath and the broad leaves of the Hoveys above; the hosts of insects and the ripening fruit around make such a residence comfortable. I know a little girl who is a great lover of fruit. She watches the strawberries. Some very large ones are taken in the hand in the evening—turned all round and carefully examined—but not being quite ripe are allowed to remain. In the morning they are gone. She asks me why? I say it is hard to know what has been in the garden in the night. Near by is a mutilated strawberry; the mark of a bite is plainly to be seen, and close by, under a broad leaf, I observe something like the eye of a Toad—a crescent-shaped streak of white just disappearing. Could that concave wound in the strawberry be brought into juxtaposition with the convex mouth of that Toad, there would probably be found a remarkable adaptation between the two; but nothing is said about it. This little girl is old enough to appreciate strawberries, but not old enough to appreciate Toads.

When I was young I was told that if I killed the Toads the cows would give bloody milk. Children in the country are fond of milk, and the fear of such a catastrophe saved the Toads.

The Toad, like the Snake, sheds his skin once a year, but the manner of doing it is very different. The snake contrives to start his skin near the head, and then by drawing himself through some tight place strips it off—skins himself—the cast-off garment being left for collectors of curiosities. The Toad works at his with his mouth, first taking off the coat, then the pants, and eats them both.

It may be asked, what has all this to do with the Curculio? Although I have often found different species of this beetle in the stomach of the Toad, the foregoing notes are more appropriate here, in connexion with the doctrine of compensation—the mixture of good and evil in this life.

May 16.—Tried my Plum trees this afternoon for Curculios, but found none. These trees are in a city garden, and have now been jarred several times. Were the young plums large enough for the puncture of the Curculio, the jarring would not so often prove unproductive; until the fruit is to a certain extent developed, they are almost as likely to be found on one tree as another. To-day, in examining the blistered leaves on some Peach trees in a neighbor's garden, I found a Curculio on the upper side of one of these blisters.

May 17.—Tried again this evening for the Curculio, but found none.

May 18.—Have seen both Pears and Cherries with the Curculio mark to-day. Found knots on young Cherry trees this afternoon, and caught a Curculio on one of them. Gave it a thorough examination under the microscope. If I had had the slightest doubt of its identity with the Curculio that is bred in the Plum and other fruits, it would now have been removed. The microscope settles all such uncertainties. By a careful examination of this knot, a crescent-shaped mark was found, with an egg in it. This egg, together with one from a punctured Apricot, was placed under the microscope, and their identity was conclusively proved. The Curculio taken from the Cherry knot was now dissected, and only two eggs were found.

May 20.—I have had quite a search for Curculios to-day, but found none upon the fruits, although a few of the earlier kinds of Pears, Plums, and Cherries are marked. Caught ten on the knots of Plum and Cherry trees; four of them on a single knot on a Cherry twig.

May 21.—The Curculios caught yesterday on the Cherry knots were taken to Mr. Hochstein to-day, that he might have an opportunity of catching the positions they assume when cutting the crescent, depositing the egg, and then securing it in the place so carefully prepared for it. Two Apricots were given them, and in less than a minute they were all on those Apricots, and the females were making the crescent-shaped marks instantly—two on one, three on the other. The males attached themselves to the stems, where they seemed to be feeding. Some of their attitudes were very amusing. Could the Elephant be photographed down to the size of one of these male Curculios, as it was attached to the stem of an Apricot, it would be very like a Curculio.

The time consumed by the female in cutting the crescent in fruit so young as it is now, is very short, not more than two minutes; but the making of the cavity in which the egg is to be stowed away, is a much more tedious operation. I waited half an hour, and none of them had finished. Many times for years past, when not so hurried as now, I have patiently watched the whole process. It is one of the exemplifications of insect instinct. The Curculio works and works at this little cave leading from the middle of the concave side of the cut in the skin of the fruit, until it attains the proper size for the easy passage of her thin-skinned and delicate egg; and at the further end of that cave or passage-way she will carefully prepare the chamber for its resting-place, larger than the passage-way, and with the adjacent pulp of the fruit so deadened that the egg will not be dangerously pressed by subsequent growth. This done, she withdraws the proboscis, or operating instrument, turns round, and drops an egg at the mouth of the cave; then turns again, and carefully pushes it to its destined place, using her proboscis for the purpose, and assuming the same position as when making the opening. If those who have seen the common woodcock boring in the soft ground for food, will carefully watch this operation of the little Curculio, they will be struck with the similarity of the positions of the two. But all is not yet finished. This crescent-shaped cut in the skin of the fruit is now carefully plastered up with a gummy deposit, of which she seems always to have the requisite supply; probably a necessary protection to prevent the separating of the wound, and the consequent exposure of the egg. It is an instinctive operation, and of course necessary and invariable.

Any one who is curious to watch all these stages of the operation to advantage, can do so by placing some Curculios in a large clean vial with some young fruit—Plum, Apple, Pear, Cherry, Apricot, Peach, or Quince—the Japan Quince, or even some of the wild berries. But I have found the following a more satisfactory way of seeing this curious procedure than viewing it through glass. Thrust the point of a knife into a young fruit, then present it to some Curculios that have been kept some hours without having a chance to deposit eggs, and they will take to it at once, giving you no trouble by trying to get away till the entire operation is completed. If your knife has a blade at each end, the point of the other blade can be pushed into some soft wood; and thus, with the Curculio at the top of the fruit on the upper blade, there will be a good chance of seeing all round.

I now jar my plum trees every day, but so far have found no Curculios, except three a week ago.

May 23, 1863. (A year ago.)—First marks on Plums and Pears. Caught three Curculios on one Green Gage tree to-day.

May 23, 1864.—Have tried three Plum trees to-day but could find no Curculios, though some Plums show marks. I now live in a city and have but a small garden.

For twelve years I had large orchards, both of Plum and Apricot trees. During all that time, except the first two years, when I was trying the various nostrums and quack Curculio remedies, I faithfully pursued the two plans recommended in this work—destroying the young fruits as they fell, so as to diminish the number of the enemy for the next year—and when the Curculio made its appearance on the young fruit, to jar—jar—jar—every day, or three times a day if necessary, till the battle was fought out and the victory won on my side. Every year of that ten years, every crop on every tree in those orchards of Plums, Apricots, and Nectarines, came to perfection. If the crop started thin I kept them all; if too abundant I let the Curculio take as much as was required for a proper thinning out, but no more.

Found several Curculios to-day on the same knots of Cherry trees and tried them with an apple of last year, but they knew it not. When portions of it were cut off and given to them, they tasted moderately.

May 25.—Found one Curculio on Green Gage tree, and two on the same Cherry knot. Killed an Oriole (Baltimore)—a male of one year; it did not have the brilliant colors of the fully matured bird. I had followed it from tree to tree for a long time, listening to its peculiar notes, and watching its habit of feeding. In a very careful examination of the contents of the stomach, what appeared to be the wing-cases of a Curculio were discovered; and on further scrutiny I found the head with the proboscis attached. This was exciting. Here was some evidence that one bird at least was feeding upon our most formidable insect enemy; but as the Curculio is one of a large family of the Coleoptera, and many of the different species bear a striking resemblance to each other, both in form and size, it was necessary to pursue the investigation still further. On placing the wing cases under the microscope, the peculiar protuberances—the brilliant metallic colors—the hairs resembling pearls, when a strong light is directed upon them, that I had so often seen, were all visible. The mutilated head was now tested. There was the proboscis with its cutting apparatus, and the 147 lenses in the eye. I have examined the eyes of many others of this family, but not one of them has the same number of lenses. The larger species figured in Pl. 5, Fig. 10, has more than double this number.

All this evidence taken together was ample to settle this question for ever. The Baltimore eats the Curculio! Let the death of this martyred bird secure the protection of its race for all future time. The remains of three other beetles and three leaf-curling caterpillars were also found in the stomach of this Oriole.

May 27.—There was rain all night, but it was over in the morning; cleared off during the day; just such weather as makes all insect life active. The great business with all insects in the last or mature stage of life is to arrange for the generation that is to succeed them. Cold and wet suspend these labors, but when the clear hot weather comes they seem to be conscious of the necessity of making up for lost time. All who have determined to protect their fruits from the Curculio must be active now.

May 28.—Caught a few Curculios to-day. Upon a close examination found the Plum crop very thin; much of the fruit stung. This must have been done between showers or one of the wet days, as the jarring has been faithfully continued, though it is certain that jarring will not invariably bring them all down, particularly on large trees. I have found them on the leaves, apparently with all their claws sticking in, as the shell of the seventeen-year locust does, as the beetle of the apple-tree borer does, after hard knocking; but when they are at work on the fruit, or moving about, as on warm days, the jarring seldom fails to bring them down—jarring, not shaking. There is a decided difference in the signification of these two words, in the Curculio business.

The wind shakes the tree, and these insects do not mind it; a bird alighting on a twig jars it, and the Curculio's instinct quickly tells it that the attraction of gravitation is its best resource from the appetite of that bird, and it falls to the ground. This any one can ascertain who has young trees just bearing. By looking carefully over the tree where the fruit shows signs of the presence of the Turk, he can easily see the Curculio at work. That tree or the branch can be bent over without disturbing it; but let it go, so that it springs back with a jerk, and off will come the Curculio. I often bend a twig so as to place fruit where I suspect one to be at work over my inverted hat, and then give it a gentle tap, and the Curculio will be in that hat instantly. A large newspaper laid down on one side, and then the tree bent over it, and tapped, answers well in the absence of the canvas. I have caught hundreds upon the New York *Evening Post.*

My Plums, I find, are half stung, many of them thrice, and some have three marks. I have spent hours to-day in taking out the eggs from the young fruit on three trees. Had the crop been a full one, I could have spared to advantage all that are now stung; but the rain that rotted the embryo Cherries within the calyx, rotted also most of the Plums, and there are none to spare for the Curculio this year. "What would Mrs. Grundy say" if the author of a book on the Curculio should have no plums!

My experience of this year in saving the fruit by the jarring process has not been so favorable as it was when the trees were smaller. There is a want of elasticity in old trees, that makes it more difficult to give the requisite jar; probably this accounts for the partial failure this year. My practice at first was the same as usual; but as soon as I saw so many of the Plums marked I came to the conclusion that merely striking large old trees with a mallet or axe was not enough, and now I use the mop-stick, as seen in Pl. 7, Fig. 5.

Caught two Curculios to-day on the same Cherry knots; also saw two Ichneumon Flies on the same knots.

May 29.—Still get two or three Curculios from the Plum trees. Try them three times a day. I had three Curculios in a small wooden pill-box. I filled the box with a saturated infusion of tobacco, and it was so tight as scarcely to leak. I have but little sympathy for Curculios. It is cruel to shut them in a pill-box at this busy season of the year; but to fill that box with such an infusion is a refinement of torture that should relieve my character of that lackadaisical reputation I have acquired with some, of being a universal friend to insects. Just at this time I was called to tea. When I returned the pill-box had exploded, and the Curculios had travelled off to parts unknown. To morrow I expect to find them when I jar the Green Gage tree. This was the strongest possible solution of the nastiest kind of tobacco. The experiment was tried over and over again. An infusion of tobacco will not kill the Curculio! It will be hard to make some people believe this; still it is so.

May 31.—Get about two Curculios a day from our three Plum trees.

June 1.—Bright, hot day. Thermometer at noon 87° in the shade, 107° in the sun. Two Curculios to-day. Have now quite a number. Dissected seven of them, five females and two males. The latter presented nothing worthy of remark; three of the females contained one egg each, although one of them was found coupled with a male; another three, and the remaining one nine. Twenty-five eggs are the largest number I have yet found. The assertion of some writers, that they deposit several eggs a day for weeks, is one of that kind of mistakes so liable to occur when people guess at things; and guessing at conclusions in the insect world is particularly hazardous.

Visited the old orchard to-day. As the orchard here spoken of is one I shall often have to allude to in the progress of this work, I will here describe it. It contains about thirty acres of land, and is now owned by a company of speculators waiting for a rise in real estate. It is about a mile from the central part of the city, and not yet wanted for building lots. There are about 300 Apple trees still standing, the remains of one of the large orchards that formerly made Newark so famous for its cider. As an orchard it is now very irregular; the trees are old, and many are dead. These trees were planted forty-five feet apart, and many are very large—measuring from six to seven feet in circumference.

This orchard is now used year after year as a cow-pasture; the fences being kept up for the income derived from the board of these cows, otherwise it would soon be a common. It borders the salt marshes, being separated from them only by a belt of swampy woods. Here great numbers of birds are found early in the season, but the idle boys of the city hunt them and rob their nests, so that they become scarce by midsummer. It was here that I shot most of the birds which I have examined, preferring to kill for scientific purposes those that were so liable to be destroyed in mere wantonness.

This has been a favorite resort for many reasons. Here I have been watching the myriads of plant lice and their effects upon the leaves and young fruit. Here I could see how closely the cows pick up the falling apples. These great Apple trees, of course, show none of the effects of the Borer, so often witnessed in younger orchards. Here I can idle away hour after hour in watching the lower orders of animated nature; and here I can shoot birds without being fined five dollars apiece. It is, in fact, a kind of Sherwood Forest for many of the younger outlaws of Newark.

Stopped to-day at the Cherry knots, but found no Curculios. The first time I have not. I now always see Ichneumon Flies more or less about these knots; they are quite small, but all of the same species. Their ovipositors are remarkably long. Find one or two Curculios a day. The Plums first stung are beginning to fall; they are so young and tender, that if placed in the hot sun they soon wilt, and the grubs die.

The Plums, where the eggs were dug out, are doing well, and will survive. Shall have a crop yet. The curled leaves of the Apple trees have generally fallen off, and the aphides that caused them have nearly disappeared; the foliage is now coming out vigorously, and is of a fine color, but the crop of Apples will be thin. Yesterday I noticed on some trees that nearly all had been stung by the Curculio. If this should be general, as I suppose it will, apples will have to come from somewhere else. Some Peach trees are well filled, but where the leaves curled badly most have fallen off, and the Curculio is at work at the remainder. Mr. P. has a fine crop of Plums; the German Prune, as usual, thin, but the black knot does not trouble it much, and the Curculio is less destructive. The Green Gage, Bolmar, and other American sorts are suffering badly, but a number of large trees of common kinds are quite full yet.

June 6.—The Green Gages, Bolmars, and some other plums are now falling, from the Curculio punctures. Those who have waited till June before attacking the Curculio, will be too late this year. Had a long conversation to-day with my neighbor Pierson about his Curculio experience. He told me that at one time he bought 100 yards of mosquito netting, and covered his young Plum trees with it; he bound cotton saturated with sweet oil round the trees, but neither did any good. Mr. Longworth, of Cincinnati, visited him, and saw his plums falling, and told him to pave under the trees. This he did, using cement to make it more complete, but after a trial of ten years he took the pavement all up as useless. Then some Yankee told him to *shake* the trees, and this was done every morning for a long time; but to use his own expression, "I got some plums nevare." Lately he has been changing most of his trees into the Quetsche or German Prune, in the belief that this variety of Plum is less liable to both the Curculio and black knot, which to some extent is true. If there were as many Green Gages or Egg Plums as the Curculios wanted, the Prunes would escape, just as Peaches would escape their attacks if they could find plenty of Nectarines.

June 8.—To-day Mr. Pierson told me more about his experiments upon the Curculio. Many of the trees still bear the mark of the *tar* with which they had been surrounded. He said that this tar injured many of them, having a binding or girdling effect, and doing no good. He said he also tried a mixture of potash, molasses, etc., "*and everything.*"

Mr. P.'s experiments were made years ago, when he was a more ambitious and enthusiastic fruit-grower than now. I have never known a more ardent amateur than he was for the first ten years of his labors. His grounds were visited and admired by all who took an interest in such pursuits. His Pears and Grapes failed from necessity—the trees and vines had been brought from Europe. All other cultivators who have tried that experiment have shared a similar fate; but the Curculio was the cause, and the only cause, of the failure of his Plum crops, and he seemed to feel that it would be a disgrace to be overcome in a contest with a little insect. This was a vastly different affair from fighting the uncongenial climate of a Continent. Then, too, the books told him precisely what to do, and with equal precision he followed their direction; supposing, of course, like most other people, that what is printed must be true. After a faithful trial, all failed to secure the desired result. Had Mr. P. taken the time to carefully investigate the nature and habits of the Curculio, or had he been able to find a treatise on the subject as elaborate as its importance demands, his good sense would have enabled him to see that all these nostrums must be useless, and he would have been saved the labor, the loss of time, and the final disgust. Speak to him now of the Curculio, and he shrugs his shoulders, but tells not what he thinks. French gentlemen will not say anything uncivil; therefore he will not say anything to one who believes the Curculio can be conquered.

If I can change all this; if I can only succeed in convincing all fruit-growers that there is no nostrum of the least value; that not one of the mixtures, washes, smokes, or smells now known will do any good, I shall have cleared the way for laying a foundation on which to build a rational system of management.

June 11.—In a visit to the old orchard to-day, it was apparent that the crop of apples was thin. The blossoms were profuse, and there had been no frost or other atmospheric cause to interfere. But the first crop of leaves had been so injured by the visitation of plant lice that most of them fell off. The ground for a few days was thickly strewn with these speckled and yellow leaves. Then soon fell also a large portion of the young fruit. But now the leaves and fruit both look well. The plant lice here, like those on the Peach trees and Maple trees that came in such vast numbers with the first bursting of the leaf-buds, are now all gone.

The apples in this orchard are less punctured by the Curculio than in any other orchard I have yet seen this year. About thirty cows are now grazing here. In other seasons the grass has been kept short all over the ground, but this year the weather has been so favorable that the growth has been greater than the cattle could consume, and the grass is now browsed close *only under the trees.* Here the cattle are attracted partly by the shade, but chiefly by the apples—the wind-falls—or more properly speaking, the apples that fall from being destroyed by the grub of the Curculio and the larva of the Apple Moth. The cows eat these apples; the embryo enemies are digested by them; and the next crop suffers less on that account. Scarcely a young Curculio will escape in this orchard. Were all the fruit establishments within ten miles as faithfully attended to, but few of the apples would fall to the ground so unseasonably next year. But the neighboring orchards are meadows, or under cultivation; the punctured fruit that falls there lies undisturbed, and the young enemy escapes, and will be on hand the next season to torment not only the owner but his neighbors.

June 12.—Found two Curculios to-day. Plums that were stung early are now falling rapidly.

We have been eating green peas from the garden since the first of the month, but I have not yet seen the Pea Bug. This beetle, called Pea-weevil—Pea-bruchus, is the *Bruchus Pisi* of Linnæus, Plate 6, Fig. 9, and is so well known to farmers and gardeners as to require little further description. This insect is somewhat similar in size and appearance to the Curculio, and some writers have said they are identical. A reference to the figures of the two, as drawn on Plate 6, will show a marked difference; the crushing test between the thumb and finger will give additional proof. The Pea-bug comes to its growth and undergoes its transformation in the substance of the pea. By opening peas containing them, late in the fall, you may see the young insects fully matured, but they will remain there till the warm weather of the next spring, unless the peas are kept in apartments artificially heated; then they will escape in the winter.

Where the great army of Pea-bugs keep themselves from the beginning of warm weather till this time (June 11th) it is difficult to say. We sometimes meet with them in places of concealment, as we do ants, spiders, flies, lady-bugs, squash-bugs, &c., &c.—torpid when cold—animated into life when warmer weather comes. We find them in the crannies of wooden buildings, fences, and walls, and there they wait their proper season. These early peas, planted in winter, are evidently too early for the Pea-bug. She is not yet ready. The Imperials, the Champions, and Marrowfats, and those sown in the fields, will come at her time. Those planted in midsummer for fall use will also escape. I have seen no account of the exact number of days that this crop is in danger from this enemy, but it is a shorter period than is occupied by the Curculio for depositing her eggs. This beetle, like the Curculio, and most others of the Coleoptera, has but one generation in a year.

The Curculio comes to maturity in the last half of July, during August and September, and some even in October. The mystery with many writers has been— Where does it live till the next May, or till the fruit comes to the proper size for it to use? and many of these writers lay great stress on this, as if it was important to be ascertained. Naturalists should know; they should know a great many other things that could be learned by patient investigation; but practically, fruit-growers have no occasion for such knowledge. They *know* that with the coming of their young fruits will come also the Curculio, unless they have destroyed it in its embryo condition in the blighted fruit the year before. Where they come from or how they pass the winter will be of little avail to prevent them from destroying the coming crops. Let *all* the young grubs of the Curculio in the blighted, wormy fruits, be destroyed while in that state, and the natural history of its winter condition will be of little consequence. Let this be done throughout a neighborhood, a township, county, or state, and the sheet and jarring process will not be so much required. The man who owns an island can, with proper care for a single year, rid himself not only of the Curculio, but of nearly all the other insect enemies of fruits and fruit trees. But ignorant, careless neighbors, who cannot be brought to terms, are a dreadful encumbrance to the man who fights these enemies single-handed. Either fruit-growers' associations, including all who own fruit trees in a neighborhood, and having laws that shall be enforced absolutely, and that will compel the necessary attention at the proper time, or total isolation of the individual, will soon become indispensable to insure success.

To-day I strolled through a fruit establishment in this State that a few years ago was quite celebrated. Great pains were taken with it for many years; more thorough and systematic manuring than I have ever known in any orchard was practised; but nothing, absolutely nothing, has ever been done to guard against the insect enemies; and a more melancholy wreck than it now is could hardly be imagined. Hogarth's "Last of all Things" would serve to tell the story of this fruit establishment. Apple trees so infested with the Borer as to be irreclaimable—mere nurseries for propagating this terrible insect throughout the neighborhood; Cherry trees becoming all knots; Peach trees full of worms; Plum trees mostly cut down—the few left, bearing profuse crops of choice kinds, but all falling to the ground when half grown. Pear trees have been grafted and re-grafted till the right kinds are at length found; and as there are few Borers in them, and little blight in New Jersey (the Curculio takes but a part of the fruit, and the Apple Moth leaves some of the remainder), the owner generally has some proceeds from his Pear crop. Had this man understood his insect enemies, and fought them resolutely from the beginning, instead of neglecting them till they became masters, he might now have had a fruit establishment not only to be proud of, but one greatly profitable, besides being a public benefactor; for he who contributes largely to the supply of wholesome ripe fruit to cities like New York, also contributes vastly to the pleasure and health of a people who are necessarily confined to such a residence in summer.

July 10.—After an interval of ten days I am again in the old orchard. The same herd of cows is still here; but, of course, the rich pasture is somewhat subdued, though much of the first growth is still standing; it is dried up, however, and not so tempting as a young crop. Under the Apple trees the grass is everywhere browsed close, and the apples are not seen lying on the ground as in unpastured orchards. Under some trees I could find five or six—seldom more—and they always show that they have recently fallen, as may be seen by the stem. Cows feed in company. In such a large range as this they cannot go all over the ground in a single day, but the apples are gathered up just as often as they do go all over; and what is more for fruit-growers, the embryo enemies within them are placed for ever *hors-de-combat*. How in a single season could every cow in this state—in every state— be made to double her value if only properly managed? How long—oh! how long— must we wait for this better management? Amateurs have improved until those fruits figured in the Frontispiece of this book are but common samples of what we could have in abundance everywhere—Apricots, Plums, Nectarines, a galaxy of luxuries— except for these insect enemies. We have the nurseries to furnish the supply; we have lists to choose from, compiled carefully by competent authority; we have elaborate instructions how to prepare the ground, to plant, to prune; but we let these little insects come without hindrance to take the greater part and deform the remainder. And this goes on year after year, of course growing worse. Could there be a year without a blossom, most people would look upon it as a misfortune. Some would interpret it as a visitation from the Almighty in punishment for our sins. Still it would be a blessing. The Curculio and Apple Moth would be checked. The former would probably prolong its race to a limited extent in the black knot; but still its ravages would be greatly diminished. Let it be known that in a single season the hogs and cattle could do an equal amount of good without the punishment of a year of privation. Let there be fruit-growing clubs everywhere, that shall make rules and enforce them, that the *fruit shall be protected from the manageable insect enemies*. Impose fines, punishments, disgrace, upon all who neglect the duty. The Hessian Fly, the Wheat Midge, and other insect enemies have compelled farmers for a time to stop the cultivation of certain crops so as to starve them out. Our hogs and cattle have both capacity and inclination to eat the Curculio out of house and home in a single season. I have been now for weeks killing poor little innocent birds, to ascertain positively what they feed on; and one object was to find which would destroy the Curculio. In one, the Baltimore Oriole, I have found the bird I sought. *The Baltimore Oriole eats the Curculio.* Probably many other birds that frequent the orchard in pursuit of food, and feed upon beetles, do the same thing; but none of them search it out exclusively. Therefore, good as most of the birds are as consumers of injurious insects, and though the world, for our purposes, would soon become topsy-turvy without them, the birds cannot be relied on to subdue or control the Curculio.

A few minutes ago I gave fifty apples to one of these cows. I had rambled over much of this orchard to find them. She ate them all in less than five minutes, and then looked up at me as Oliver Twist looked up at Bumble, and almost as plainly said, "more."

July 11.—Examined the Curculios to-day. The earth in the flower-pots and boxes in which I have been throwing the fallen plums, is teeming with the grubs. A few begin to show embryo legs and wings; but none are yet as far advanced as Figs. 4 and 5, Plate 6. Most are still larvæ, and seem to be busy making their cells. Nearly all the plums are now empty, showing the holes, as in Figs. 5 and 6, Plate 3, where they have gone out. Have placed gauze over them to prevent escape. Find no Curculios on my own trees. I notice that Green Gages and Bolmar plums, now falling from the puncture of the Curculio, show specks of *rot*. This certainly is not caused by *wet weather*.

July 12.—Tried my Plum trees this morning, but found no Curculios. Took the canvas and mop-stick to Mr. P.'s, and got sixteen from four trees. Would have tried other trees, but had only two pill-boxes; and owing to the heat, the Curculios were so quick in their motions, I could manage but few in each. Two did not always escape when I put one in, but one was pretty sure to get away when I tried to put two in.

I am not an entomologist, and never expect to be. If I knew all about all the insects, I would be willing to accept the title. The fact is, I do not believe I know all about any one insect. Here I have been watching this thing twenty years. I see it come into the mature or winged condition in the summer and fall. The next spring, in May, it will be depositing eggs in fruit. I see it still in the same condition in the middle of July. The Pea bug, which somewhat resembles it in appearance, runs a similar career; but most other insects pass much the largest period of their lives in the larva state. The Cicada, for instance, is sixteen years, nine months, and ten days in the earth, and about twenty days above ground in its perfected form. Some varieties of ephemera will be four years under water, and perhaps only that many hours in the winged state, in the air. How mysterious—how wonderful is this little insect world! I am now watching these sixteen Curculios in a glass covered with gauze. They have two plums, and a few of them creep over and examine them, especially the specks of gum; but the greater number are restless, and try to get out. The last time, about two weeks ago, that I examined the operations of a company of the Curculios on fruit, several of them passed a long time cutting the crescent marks; but I could not see them deposit the egg. I now wish to test this matter further, with reference to the rot in plums about this time.

Mr. Pierson's man has been gathering the blighted apples that lay in the walks, but only because they were a deformity to the garden. They were thrown upon the manure heap, where the Curculios will probably have just as good a chance to come to maturity for next year's mischief as if they had been left on the ground. Mr. P. has hogs, cattle, and horses, that would have been glad to eat them, worms and all, but he has been so disgusted with his experience in fighting the insect enemies, that he will hardly listen kindly to further advice. He probably feels as Job did towards his comforters: "Ye are forgers of lies, ye are all physicians of no value." But to-day, finding he had done such a foolish thing, instead of following my advice, I gave it to him sharply. I gathered up the apples under the half of one tree, where they had not fallen on the walk. I was just sixteen minutes in doing it, and picked them clean, too, and there were 960; that would be 1,920 for the tree. Probably hundreds more will fall, some from the effects of the Curculio, some from the Apple Moth. In many of these apples the larvæ both of the Apple Moth and Curculio had escaped, especially the latter. In some I counted three and four holes. Of course the fallen fruit should have been gathered sooner than this; even apples should not be permitted to lie on the ground later than the last week in June. Let us carry this further. All the fallen fruit under an Apple tree, to the number of 2,000, can be gathered by hand in about half an hour—twenty trees in a day. If your trees are so situated that the hogs, or cattle, or horses, or sheep, cannot do it for you, have it done by hand. Do it yourself, if possible, but have it done, and well done. Children that can do but little else can attend to this. Stimulate them with a fixed pecuniary reward for every bushel. No money could be better invested. Let a neighborhood do it, and what a difference in the fruit crop the next year! Hundreds of people ask, "Is there any cure for the Curculio?" I answer, yes; this!

July 14.—My sixteen Curculios from Mr. P.'s are still alive. The two plums given to them at the time are now punctured in at least one hundred and fifty places; cut with the end of the proboscis, as usual. Eggs were found in about a dozen of these crescent cuts. In some the insects had eaten cavities, as if for food, but in most there was only the crescent mark, though in others they had also made the centre opening, as if destined for the egg.

I examined my box of young Curculios to-day, and found that eleven had come up out of the ground fully matured, the colors as dark as those of the old ones, but of much brighter tints than the parents are in their old age. Here is another complication. Two generations of Curculios at the same time! One is trouble enough certainly, but I shall try to see and record the doings of both. I have placed this new brood in a glass vessel with a gauze covering, and given them plums and pears.

A near neighbor has an Apricot tree that stands against his house. I have watched it for years, and I have not seen a mark of a Curculio on any of the fruit. He has also many Plum, Peach, and other fruit trees in his yard, but none of them so near the house as the Apricot tree. All these suffer severely from the Curculio. Instances of such exemption from Curculio depredations have often come to my knowledge. I had such on my own place. This neighbor has many common Plum trees, standing where they came up as sprouts; common sorts, proper subjects for the black knots. These knots are now perforated in all parts, though generally from one end to the other in the centre, by grubs of Curculio. The yellow-brown powder is issuing from openings all over these knots. Many of these have now come out, as from plums or apples, and gone into the ground. Let any one who doubts this try the experiment; it is easily done. Earth in flower-pots will do; and if covered up with gauze, the beetles can be secured for examination. Those bred from the plums and plum knots will be found identical. In this I do not wish to be understood to say that it is the Curculio that causes the black knots on Plum and Cherry trees, but I do intend to be understood to say that the Curculio is bred in many of the black knots on both of these trees. Some years ago one of the agricultural papers gave an account, from a correspondent in Canada, of an Ichneumon Fly breeding in the Curculio; but it was in the grub as found in the black knot.

July 15.—The young Curculios I find nibble at a pear, but cut no crescent marks, and make no holes with the proboscis, as the old females do. *I dissected eight of them, but found neither eggs nor signs of them.* I then examined three old ones, and in one found a single egg, and on a careful examination of a plum, which these three had had for two days, I found three eggs more. This having two generations of Curculios on hand at the same time is not so complicated an affair as it appears to be at first. These experiments prove that those of this year are not so far matured as to be ready for the great purpose of their lives—the propagation of their species. Female insects which pass but a short time in the winged or perfect condition, as moths, butterflies, cicadas, etc., will be found to contain eggs at the time of emerging from the chrysalis, and the sexes come together almost immediately. But with these young Curculios the case is different. Some that have been kept in a green-house until mid-winter, as they have been by Peter B. Mead, myself, and some others I have heard of, will be found pairing, but this must be considered as the effect of the artificial temperature. Out of doors they will be in the condition of others of their class, as lady-bugs, asparagus beetles, and pea-bugs, too nearly dormant at all times in the winter to show much activity; often they will be frozen solid. Occasionally there will be a day so warm that many of these insects will be noticed peeping out, but there will be no signs of intercourse between the sexes.

That degree of temperature that draws forth the bud and the blossom the next spring, brings these Curculios to that stage of life that their eggs will be ready for the young fruit when the young fruit will be ready for the eggs. The idea of some writers, that there are two generations of this insect that prey upon the fruits the same year—one generation early in the season, another later—has been proved to be erroneous. Two generations there are undoubtedly living at the same time, but I have only been able to find the egg in the female of the older generation.

July 16.—The last time I was at the orchard of my friend, Mr. P., jarring some of his Plum trees for Curculios, I tried also the experiment how many of the little apples I could pick in a given time. I had put about a bushel of these blighted apples in my canvas, and tied them up so that I could see on my next visit how many of the grubs of the Curculio and larvæ of the Apple Moth would have come to their growth, and escaped from the fruit. But the sheet had been seen lying there, and was supposed to be the cover of the carriage. It was ordered to be taken up, and the apples, worms, and all, being emptied upon the ground, my experiment was spoilt. These apples had been emptied in a heap, and we gathered them up carefully; the little spots of brown powder that had fallen out where the worms had made the holes to escape from the apples were plainly to be seen. Many of the openings could also be detected where the grubs had gone into the ground. Some I found about an inch under the surface. But the ground here, from the excessive drought, was too hard to dig much with a pocket knife (and I had nothing else), and hundreds of my friend's enemies will probably come up from this spot to torment him next year, unless this dry weather should last a few weeks longer.

July 17.—I have been jarring four of Mr. Pierson's Plum trees to-day, and caught nine Curculios. The jarring brought down a great number of plums. I counted 160 that fell from one side of one tree, and *all* the effect of the Curculio.

July 18.—The nine Curculios caught yesterday have been kept in a bottle with two Plums (the Quetsche). They have left many marks, and deposited a few eggs, but this kind of plum seems too hard to suit them. The Green Gage or Bolmar would have shown more marks in the same time.

July 20.—On a tour of observation to Western New York. In a ramble in the outskirts of Buffalo and Black Rock, I found some fruit trees in the pasture lots and gardens. The black knots were on the Plum trees, but I saw none on the Cherry trees, as in New Jersey and many other places. In this ramble of several hours I examined all the Apple trees that I could approach without seeming to be too curious about the gardens of strangers. I saw no marks of the apples having been punctured by the Curculio. A few showed signs of the Apple Moth; but apples so fair I had not seen for many years. It was truly refreshing. I had just come from the orchards of New Jersey and Eastern New York, where half the apples had already fallen, and most of the remainder were blighted and deformed by the Curculio, and I was amazed at such a vision. It was a new sensation. Whether this is so every year in this locality, or whether there had been a drought the preceding year just at the right time, I was unable to find out. At 3 P. M. I started in an omnibus for Williamsville, ten miles east of Buffalo, and two or three miles further, on foot, brought me to the home of the parents of my friend Anthony Hochstein, the artist, to whom this book is indebted for its artistic merits. The apple orchards here were, like the few trees near Black Rock, well loaded with fruit, and there were no signs of the Curculio. I was told that the plums were entirely destroyed every year by the Curculio, but I saw none; a few peaches in the garden here were badly punctured.

July 21.—In returning to Buffalo I preceded the stage on foot several miles to examine the orchards along the road, and found scarcely any Curculio marks. The Apple Moth had been at work, though not badly. There were plenty of the tent caterpillars. In the afternoon, took a long stroll about the city. Saw one Plum tree full of fruit, but it stood close to a house. Bough apples were on the stands in the shops. They were badly marked by the Curculio, but they had come from further South, probably from Cleveland, Ohio. The season at Buffalo is two weeks later than at Newark, New Jersey. Cherries and Doolittle Black Caps were now in perfection here, but they were all gone many days before I left home.

July 23.—The nine Curculios caught last I found dead when I returned. The Bartlett Pear given them was untouched.

July 28.—Took a trip down the New York bay to enjoy the sea breeze during the heat of the day. In a ramble on the eastern end of Staten Island I examined the trees as far as possible, but most of them were without fruit, and showed signs of rough usage. A man with an excitable temper should not live near a steamboat or railroad landing, or, if he does, should not attempt raising fruit. Those who do not believe in the doctrine of original sin can know very little about boys, who, it was plainly to be seen, had been at work here. Some boys would be as useful as pigs or cows in destroying defective fruit, but others are good for nothing, as they spit out the worms.

July 31.—I have been examining the colonies of young Curculios. Those from the plums are nearly all perfected; that is, the grubs have become beetles. The worm, as we call it, as met with in the fruit, has become a winged insect, as unlike its former self as a bud would be unlike a blossom, or a tadpole unlike a frog. About one in ten were still under the ground, their transformation not yet completed. One only had not yet begun to change, but had made its cell, and was waiting. The apples are producing very few Curculios in proportion to the number stung, which appeared to have fallen prematurely from that cause.

The Curculio, during the fall, feeds on both leaves and fruits. It will make plums or soft apples or pears cellular with little excavations into the pulp. Often it will be found stuck fast and dead in the fruit that has rotted.

The Curculio requires plenty of air. It will soon die in a box or vial if the air is excluded; but I have kept them for weeks in pint or quart glass bottles with large mouths, if covered with millinet, or corks with holes in. In their efforts to escape they will often work through both millinet and cork. I have known them nibble the latter till it became perfectly cellular, and they were covered with the dust, as a bee often is with pollen.

Aug. 1.—In a walk in one of our quiet streets, where most of the houses have neat little gardens attached, I noticed two Apricot trees with a fair crop of fruit, and quite ripe. Apricots this year are small, owing to the excessive drought, and these were a small variety, but very pretty, the rich yellow giving them a tempting look. Moorparks would have been three times the size. These Apricot trees were as large as old Plum trees. There are few more elegant ornaments of the home than such trees when loaded with ripe Apricots; which besides giving plenty for the family, enable us to send a basketful to an invalid friend, or the wounded soldier in a military hospital. This world is not yet quite all bad, and it can soon be improved if we go to work resolutely and save the fruits. Reader, take another look at the Apricot in the Frontispiece, and then say, *the Curculio shall no longer stand between thee and me*, and let all the neighbors say Amen.

In a further walk to-day through several of the streets of this city, where well-to-do mechanics live, I noticed that most of the Plum trees bore some fruit, and some were quite full. These trees are generally old and large, but being in small gardens, and planted too near together, grow tall. Many were near buildings; those with most fruit on were very near white houses. The principal varieties were Bolmar's, Yellow Egg, Green and Blue Gages, Golden Drop, and some of more common sorts. Probably none of the newer kinds are planted. The people have so universally yielded to the dreaded Curculio, that the raising of Plum trees in the nurseries in this part of the country has been abandoned.

In a visit to Mr. P.'s to-day, I jarred four of his Plum trees over a large sheet, but caught only one Curculio; probably the last of the season, but its terrible effects still remained. Down came the plums by dozens. These were all superior sorts, and now nearly as large as when ripe. In another month they would have been very tempting. Upon a careful examination of a great number, I could find the crescent mark in all, but not a single grub. Many were quite rotten, some half, some only a speck; and some had fallen without any sign of decay, as they do earlier in the season. Many people attribute the rotting of plums just before the period of ripening, to the weather. This year, if the weather was the cause, it was too dry. Last year, as it rained all the time, it must have been too wet. The weather is always a convenient resource to those who do not look for causes beyond it. With some, be it hot or cold, or wet or dry, it is all the same. The weather, undoubtedly, has an influence upon delicate fruits when nearly ripe. Cherries, apricots, plums, and even some pears, will spoil rapidly, if the weather should be wet, foggy, and hot (what is usually known as dog-day weather), at this critical time. But now, a month before the plums are ripe, no weather will cause them to rot, unless there is a wound. I have known the striking of hail-stones on plums make a bruise sufficient to cause them to rot. But there seems to be a poison from the punctures of the Curculio late in the season that is peculiarly fatal; what it is, it would be difficult to say. It is certain that the egg is not often hatched when deposited in fruit when the pit is maturing. This may be observed not only in the plum, but in the pear, nectarine, and apricot. Many such fall, more or less decayed; others will remain upon the tree and dry up (see Plate 1, Figure 5, and Plate 5, Figure 6). Early in the season, in the plum, nectarine, and apricot, the egg is nearly always hatched, and the young grub comes to maturity. I have sometimes thought that the acid juices of the maturing fruits may interfere, but I have no proof, and my readers must not mistake a mere suggestion for a theory. I have had too much experience to indulge in any theories on insect operations.

The plums that fell to-day on jarring the trees had been punctured by Curculios, and most of them were more or less rotted. Those without Curculio marks were sound, and stuck fast in spite of the jarring. Wet weather could not have been the cause, for we have had but one rain for six weeks, and that lasted only half a day. It has been very warm at times; so it is every year. You wicked ones, who are so prone to ascribe your misfortunes to the weather, refrain from all such croakings. The weather is regulated with a wisdom far beyond man's comprehension; cease to blame it, and look to other causes. Your own want of knowledge, or your own neglect, will often account for your troubles. When your Plums rot just as you think they are safe and almost ready for the market, ascribe it to the Curculio and not to the weather.

Ten days ago I left a bag at Mr. P.'s to be filled with the falling apples. To-day I examined it. The bag was open. I had requested it to be tied up when filled; but John's knowledge of English being poor, and mine of French not much better, the young enemies had a chance to escape. In this bag I found thirty-one worms of the Apple Moth. Most of them had formed their cocoons, attaching them to the inside of the bag. Some were still among the apples. How many I should have found had the bag been tied cannot be known. Only three grubs of the Curculio were to be seen; but there was a hole in the bottom of the bag that one had made, and many may have escaped there. This experience was the reverse of what might have been expected, certainly different from that of other years.

On examining these apples scarcely one could be found that had not been punctured by the Curculio, and many have more than one mark. Cutting these little apples into pieces I readily saw the minute roads which the young grubs had made, as in Pl. 5, Fig. 4, but all had come to an untimely end before they had reached such a size as to be plainly visible to the naked eye. In other seasons I have found the grubs in such proportions to the apples, in experiments of this kind, that a double handful might have been gathered from such a bagful. And this has been my experience nearly always for twenty years till this season. What had brought them to this sudden end this year I do not know. The influence of the weather on some insects is well known. That a drought of some duration during the period of transformation in the ground is fatal to the Curculio, I have proved again and again. I suppose the earlier drought of this season killed them in the fruits; but I have no proof to offer, and it is only an opinion. We have had two days this season when the mercury in the thermometer indicated 100° in the shade. This intense heat may have killed these young grubs even in the apples; but of this I am not certain.

Aug. 4.—CROOKED LAKE, YATES CO., N. Y.—The Peach trees here have a very thin crop this year, and most of the fruit on those I have been able to examine is badly stung by the Curculio. The crop of apples is better, but it also is badly injured by this enemy.

Aug. 5.—This part of the country is devoted to wool-growing. Merino sheep abound. I encountered a company of some twenty bucks in an orchard. They are queer-looking animals. Imagine a Broadway exquisite and one of these sheep brought face to face, and the former to be told that the coat that makes him about all he is, had been worn a whole year by the latter, before it had fallen to his portion—that the man is the shoddy of the sheep! What man owes to this uncouth Merino woman owes to a repulsive caterpillar; and how little credit either ever gives to the real producer of what they are so proud of. " Such is life."

I have been called the " Curculio man." I certainly have tried to find out how most successfully to secure our fruits from the depredations of this insect. Some men write long articles about the comparative values of the domestic animals, but the Curculio-destroying merit has seldom entered into these calculations. Those who have Merino bucks valued at a thousand dollars a head, may, after the experience of to-day, add a good many more dollars to this thousand. No apples were to be seen lying on the ground here. I picked some green and hard from the trees, and the sheep ate them with avidity. One of them amused me by his repeated attempts to help himself from a pendent branch; he could just touch the apple with his nose, when standing straight up on his hind feet, but the fruit would slip away as he attempted to take it. The part he could reach was a segment of too large a circle for that narrow mouth. He tried again and again; the seventh attempt was successful. " Patience and perseverance overcome difficulties." Sheep may be added to the other domestic animals qualified to settle the Curculio question.

Aug. 6.—The approach to Rochester from the east shows to what an immense extent the nursery business is carried on, and a visit to Ellwanger and Barry proves that it must be profitable, at least in one instance. I was in pursuit of an apricot for illustrating this book—a Moorpark, such as I have grown by hundreds of bushels on the Upper Hudson, but could find none—a few trees, but no fruit. The Plum orchard in this establishment is superb, and well loaded with fruit. Here can be seen most of the kinds now known; and how any one who has seen such an orchard can resist the temptation to have one for himself, is marvellous.—Let us hope that this magnificent fruit, now so neglected, will soon be restored. These gentlemen have it every year. The Curculio is in Rochester, and would take their plums, but they do not let it. Others can do the same. There is no mystery about it.

They have no nostrum, no Curculio cure, and they know there is none. There is prevention; but if it has not been practised, and the Curculio comes upon the young plums, they are jarred down upon canvas and killed.

The Jaune Hative was now ripe, and looked tempting; but, like the earliest Peaches or Pears, not so fine as later kinds. Some varieties were rotting, and where they grew in clusters were affecting all they touched. It was easy to see that the puncture of the Curculio had been the cause. The foreman remarked that the jarring had not been continued long enough.

Aug. 7.—NIAGARA FALLS.—Rambled round Goat Island before breakfast. It contains sixty-two and a half acres, chiefly the original forest. Great trees of bass wood, elm, maple, some that had trembled every moment for centuries from the pouring of this mighty torrent of water. Near the bridge is a house and garden. A few fruit trees were growing here. The apples were very badly deformed by both the Curculio and Apple Moth. Peaches also suffered from the former. A tree of Sweet Bough apples at Suspension Bridge was examined. Half the fruit was spoilt by these two enemies.

Aug. 10.—Home again, and have been examining some of my insect propagating houses to-day. The Curculios from plums have all escaped, having forced openings through the millinet or gauze coverings. The Curculio is hard to manage, but wire gauze would have prevented this. Those from the bushel of apples from Mr. P.'s are now in the ground, undergoing their change; but I see very few of them, considering the number of apples, and the proportion of other years.

Aug. 11.—Much has been written about catching injurious insects in widemouthed bottles partly filled with sweetened fluids. Some have said that they have caught the Curculio in that way. I have an impression that a great many people do not know the Curculio when they see it. For some days I kept a bottle thus prepared, hanging in a tree in the garden. When I returned from my recent trip of a few days it was full. It was bridged, so that more insects could go in and not be drowned. It is said that there are ants in South America that, when they emigrate, move in a straight line, never turning aside, no matter what may be the obstruction. If they encounter a stream of water they rush right in, until it is so bridged with their bodies that the rest can go over with safety. So it was with this trap. I took out the insects, separated and counted them. There were 571 flies of eight different species, 8 small moths, 3 mosquitoes, 1 lace-wing fly, and 1 wasp.

Not one of the white moths that are now so numerous—the mature insect from the army of caterpillars that made such sad havoc with the foliage of our fruit trees a month ago. The flies I have not yet examined closely. Many of them are the blue and green bottle-flies, the maggots of which are the consumers of decaying animals. A very large proportion were those flies that extend their wings out at right angles from the body, and have brown heads and bodies. These breed in and consume the filth about our houses and barns. I noticed some varieties of the sulphus-fly that deposits its eggs in colonies of plant lice, on which the young maggot feeds, and thus befriends the gardener. Many also of the parasitic class, that deposit their eggs on the sides of large caterpillars, near the head. The young from the eggs eat into these caterpillars, and, there feed upon their living flesh, finally destroying them; and these also are often our friends. The lace-wing is also an enemy to the plant lice. The wasp's reputation is equivocal. He destroys insects, but will also eat our ripe fruits.

In order to make a more exact test, I repeated this experiment for twenty-four hours. The number of victims was 281. The kinds of insects were about the same, and nearly in the same proportions as at first. To this statement I find appended in my diary the following observations: "This is a poor business, and I shall stop it. What right have I to destroy these hundreds of flies, when we are certain that most of them are useful, and we do not know that any of them are injurious! If I had found the Curculio, Apple Moth, Aphides, Pear or Apple Tree Borers, or Mosquitoes in any proportion to the vast army of this pest now filling the air; had the flies been of the kind that annoy us in the house, or had I caught the moths of the caterpillars that have been so troublesome, there would have been some excuse. But this is like firing into a crowd of friends in hopes of killing an enemy."

Aug. 13, 14, and 15.—Am passing a few days in the northern part of New Jersey, to escape the distressing heat of the city. I spend hours in the old orchards. Have been cutting the blighted apples till knife and hands are black and sticky with the juice. Find no grubs of the Curculio, but thousands of their marks. The crescents had been made, the eggs deposited, most of them had been hatched, and the young Curculio had started on its rambles towards the centre of the fruit. Its pathway could be traced by a brown or green mark, as seen in Plate 5, Figure 4. This mark is not visible when first made; but, like a wound in the flesh of the apple from any cause, soon becomes discolored. My experience here was similar to that in Mr. P.'s orchard in Newark. The grubs of the Curculio were nearly all dead.

Aug. 16.—I have seen an old apple orchard to-day so different from others in the neighborhood of the same age, that I visited the owner to find out the cause. He told me that thirty years ago the ground in this one had had a coat of lime, the others none, and that this was the cause of the difference; but he could not tell me why the fruit was so much fairer and the crop larger. I asked him if it had generally been used as pasture ground for horses, cows, and hogs, as it was then. He said it had.

Aug. 17.—While detained at the neighboring village of Dover, waiting for the train, I inspected an old apple orchard with broken fences, and where the cattle and other animals grazed without restraint. The fruit here was like that seen yesterday, quite fair. These are suggestive facts.

Aug. 19.—Yesterday was warm and wet, such weather as hurries on the ripening of the fruit. Green Gages, with dry weather, will last in perfection a week or ten days; but now many of them are cracked, and the wasps and flies find how good they are. The period of nectar that these fine old plums afford will be short this year. After the hard fight with the Curculio it is a disappointment to be able to enjoy them so short a time. In England and other parts of Europe where they have no Curculio, most people know practically what the Green Gage is; but here the knowledge is confined to the older generation, and with most people it is only a tradition. Almost every one has a standard of excellence in fruit. I have seen a catalogue of the great nursery at Angers in France, in which the Seckel pear was pronounced the best in the world. I have eaten this pear from the original tree in Kingsessing, Pennsylvania, and from other trees equally good. Were it not for the Green Gage I should say that it is not only the best Pear but the best fruit in the world. But though other Pears are not equal in excellence to the Seckel, yet in other respects many have great advantages over it. The same may be said of Plums. There is one that many would pronounce equal to the Green Gage in flavor, and the fruit will ripen in succession for several weeks, and still I cannot find this tree in any of the nurseries I visit. I allude to the Mellen Gage, a native of Hudson, N. Y. I could name thirty other varieties of great excellence, either for flavor and size of fruit or for bearing qualities.

Aug. 23.—Have tried Mr. P.'s Plum trees again for the Curculio and found none, either old or young; but how the rotten plums did come down! Several of these trees were the Flushing Gage, and had been very full. What bushels and bushels this gentleman would have had with proper care!

Sept. 1.—Mr. P. told me that he had gathered his plums to-day, "Had thirty in all instead of thirty bushels." The ground under the trees presents a sad appearance. I have seen, where over-ripe persimmons have fallen from the trees in winter, the ground covered with a yellow mash. The ground under the Plum trees presented a somewhat similar appearance. Many were still on the trees, looking like those in the Frontispiece, Fig. 5.

Sept. 10.—NEW VERNON, MORRIS CO., N. J.—Here are Peach orchards celebrated for the excellence of the fruit, and especially for its beauty of color. The best trees as well as the best fruit grow where the soil is strongly marked with the presence of iron. The owners of the principal orchards here were from Monmouth Co., N. J. They understood the cultivation of the Peach thoroughly, and were making the business profitable. They told me that they considered the Yellows a good thing; without it peaches would be so plentiful as not to pay; as it is, good cultivators can make money. The Curculio they found troublesome in thin crops, but it does good to heavy ones. These men might be called Peach Philosophers; they seemed to be satisfied with everything excepting only the war; that preventing them from getting Peach pits from a part of Virginia where the trees are still free from the Yellows. The favorite peaches here were Jaques or Yellow Rare Ripe, Mountain Rose, Stump the World, Crawford's Early and Crawford's Late, and Keyport White.

Sept. 13–15.—United States Pomological Convention at Rochester.—I shall long remember this meeting with pleasure. The characters of men devoted to Pomology are probably influenced by the pursuit. The intellectual man is moulded into shape by the beauties of nature, and he becomes intelligent and good. A very small percentage of the whole people of this country are members of the Society of Friends, but a very large percentage of the members of this Convention were of that denomination.

Dec. 5.—I find no notes in my diary about the Curculio for nearly three months. I see its marks on the apples every day. No matter where I go or where the apples have come from, there the Curculio has been. The signs of its former presence on many of these are generally in the form of a little shield, as seen in Plate 5, Figure 1, a mere discoloration, without causing any injury—not even a deformity. In many the mark will be little more than a slight depression. In some there will be several of these, and the apple will be much injured both externally and internally, as in Plate 5, Figure 3. All this may be seen in the fruit that comes to the markets; but go to the orchards or cider mills, and it is ten times worse. With all this, becoming worse every year, many people, when told that they must promptly destroy the falling fruit, will say, " That is too much trouble. I cannot have the cattle or hogs in my orchards, I want to plough them or mow them." Or, " If I should do this, my neighbors will not, and what's the use ?" Well, go on then " in your old shiftless way," and those who do take the proper care will have the greater profits.

I expected to have had a colony of Curculios for investigation this winter. I had prepared a frame covered with wire gauze in the shape of an old-fashioned wire trap, about two feet in diameter. This was to be placed on the ground under an Apple tree, and the Curculios from a bushel of apples to be kept under it; but the bushel of apples produced but eight instead of a thousand, as I expected. The eight I found dead when I returned from one of my trips, and it was too late then to repeat the experiment this year.

Whether the Curculio, after undergoing its transformation in the ground as it always does, and then coming to the surface, ever goes back again as winter approaches, I do not know. Once, in repairing the roof of a house late in the fall, I observed some of these insects in a torpid condition under the shingles. I have met them in the chinks of stone walls, and once I found one under a scale of bark of an apple tree near the ground early in the spring.

A favorite practice of the writer for years, has been searching under the rough bark of different kinds of trees, to see what species of insects find shelter there, and what condition they are in during different degrees of temperature. Lady-bugs, asparagus beetles, spiders, flies, great varieties of beetles, and ants are there, but I have not found the Curculio except in the one instance. Peter B. Mead, Mr. C. Marié, and I, have kept them in green-houses in the winter, and found they will feed on green leaves and fruits when warm, and the sexes will sometimes be together; but when colder, all become quiet and seem torpid, though a momentary warming, as under the thumb in the palm of the hand, awakes them into activity at once.

Many believe that the Curculio lives through the winter in the immature condition of the grub, and undergoes its transformation in the spring. This is not so. In all my numerous experiments made year after year, even with the latest stung apples, the grubs become beetles the same season, and as beetles they live somewhere through the winter. When the warm weather of the next spring is fairly established, and before the young fruits are formed, a few Curculios can occasionally be found by jarring almost any kind of tree over canvas, provided that tree is in the neighborhood of where fruit grew the year before.

Dec. 12.—In a conversation to-day with Mr. Marié, about the Curculio, he told me that he had kept some as pets until late in November, feeding them with pieces of apples. Finally they were thrown out accidentally, and swept into the fire. I have often kept them in the flower-pots in which they have undergone their transformation, by covering them with millinet. Some try to escape through this covering, but most of them will be found curled up and torpid, lying among the little clods of earth. During the warm weather of early fall they will nibble a little at fruit or leaves, but as the cold weather approaches they become quite torpid.

Mr. Marié's experience has been like my own, as to the great tenacity of life of this insect. He has experimented with many things, with reference to finding something that might be made available towards its control. He told me to-day that he had tried them in the strongest mustard, all kinds of pepper—even the hottest cayenne, French vinegar, and chloride of lime. They not only survive all these, but soon appear as lively as ever. Tobacco smoke, puffed upon them, if long continued, will kill them. Oil is fatal to them, as it is to all insects.

The following short communication appeared in *Moore's Rural New Yorker*, of Jan. 28th, 1865:—

"*How to Catch Curculios.*—Eds. Rural New Yorker: In May last we had occasion to use some lumber. It was laid down in the vicinity of the Plum yard, and on taking up a piece of it one cold morning, we discovered a number of Curculios huddled together on the under side. On examining other boards we found more, so we spread it out to see if we could catch more, and we continued to find more or less every day, for two weeks. We caught in all one hundred and sixty-one. So I think if people would take a little pains they might destroy a great many such pests. These were caught before the plum trees were in flower. What is most singular is, that we never found a Curculio on a piece of old lumber, although we put several pieces down to try them. They seemed to come out of the ground, as we could find them several times a day by turning over the boards,

"*Johnsonville, N. Y.*, 1865. Mrs. H. Wier."

Upon which the Editors remark:—

"These facts are interesting. Observers do not agree as to whether the Curculio remains in the ground during the winter or not. Some assert that it lives above ground somewhere in its perfect state or form. Any facts relating to the settlement of this question will be interesting."

The above struck me as containing important facts, provided the insects found were really the Curculio, and I at once wrote to this lady to send me some specimens, if she had kept any. In a few days I received an answer, stating that they had killed them all at the time, but would certainly try the experiment in the spring again, and send some then. She stated that she was perfectly familiar with the Curculio, from having so often seen it at work on the Plums, and she had no doubt that those found collected in clusters under the boards were the same. I immediately wrote again, inclosing a few dry specimens in a quill, and a part of Plate 8 of this book. The following is her answer:—

"JOHNSONVILLE, *March* 1, 1865.

"Yours of the 27th was received last night, and I hasten to reply. The Curculios you send, we consider identical with those we caught under the boards; although dry and contracted we can see how they ought to look. The one in the drawing looks very natural, but they do not always wait till the fruit is so large here, before they begin the work of destruction.

"I will keep these specimens and the drawing for further comparison.

"MRS. HENRY WIER,
"Johnsonville P. O., Pittstown,
"Rensselaer County, New York."

Mrs. W. states in one of her letters that she has no record of the time when these insects were found, but that the trellises for the grape vines were put up on the 12th of May, and that these boards had been drawn for that purpose and laid under a plum tree near by, a few days before. If the printers now at work setting the types of this volume would wait patiently another month, I think this question of the winter condition of the Curculio could be ascertained; but as this cannot be done, I will here indulge for once in a theory—a speculation: That this insect remains through the fall, winter, and early spring, very near the surface of the ground, in the little cracks or fissures where bare, and about the roots of the grass where in sod. If this should prove true, it may lead to some new modes of treatment; though I cannot imagine at present anything that would supersede the plan of thoroughly destroying all the young fruit containing the embryo insect.

Poultry have been proved to be useful. If these little beetles lie at all in sight of their sharp eyes during the winter, keep them as much as possible among the fruit trees, provided the crop had been troubled with the Curculio the season before. There is much weather of every winter when the ground is bare of snow that the poultry will be found searching the fields and meadows for insects. *If the Curculio is within reach*, let the sharp eyes of the poultry have a chance.

In February a year ago my friend John T. Hicks, of Westbury, Long Island, N. Y., showed me a box containing the contents of the stomach of a crow that had been shot a few days before. The box contained a few beetles, and about fifty grasshoppers. Some of these were of the variety so plentiful late in the fall, but the greater part were of that kind that we find in the spring about half grown, and not yet having their wings matured—such as are at full size in July. Many do not know that grasshoppers live through the winter; many do not know that crows eat insects. The farmers, when they see flocks of crows ransacking their fields and meadows, instead of offering bounties for their destruction, should be thankful that there is something to keep the grasshoppers and other insects in check.

In April last I dissected a Meadow Lark. I found its stomach filled with the rings or sections of what are often called thousand-legged worms—Iules. These insects are found about the roots of grass a little under the surface. The beak of the Lark is long, strong, and tapering to a very sharp point. The beak of this one was coated with earth.

In July I shot another. It had been feeding on beetles and other insects commonly found on the ground in pastures. In December, when the ground had been covered with snow for several days, a Lark was opened. It had been feeding on seeds, chiefly oats and wheat. Crows shot at this time were found to contain black marsh mud, with here and there the off-shoots of bulbous roots.

The winter birds as well as the poultry, except when the ground is covered with snow, will help us in the destruction of our insect enemies. The Grasshoppers in the stomach of the Crow, and the Iules found in the Lark, show that they have a faculty of detecting the hiding-places of these insects in their torpid condition, or a sense of sight wonderfully acute. It is a recognised fact that the Crow knows from the wilted appearance of certain plants where to find the destroying grub at the root. Could we know exactly what the stomachs of a party of hens contain after a foraging expedition to an orchard, we should probably re-arrange our figures giving the profit and loss account of Poultry.

"*Protection against the Curculio.*—It has frequently been remarked that fowls were more or less a protection against the Curculio. A striking example of this has been shown the present season in the grounds of Wm. H. Southwick, New Baltimore, N. Y. He has many very handsome Plum trees, of good size, healthy, and vigorous. Several of these trees of different kinds are inclosed in yards where fowls are kept—separate inclosures being necessary for the different breeds which are here bred. The trees in the fowl-yards are loaded with plums, while on the trees not so inclosed almost all the fruit has been lost by the sting of the Curculio."—*Cultivator*, Sept. 1861.

NOTE.—On page 49 of this volume, on the next line to the bottom, the word "*peach*" is printed instead of "*pear.*" This is an error of so much importance as to require an explicit correction.

The young Pear crop will be much benefited by jarring off the Curculio for a few days. After this the Pears will be deserted for other fruits; but if the Peach crop is to be protected from the Curculio the jarring should be continued as with Plums or Apricots.

UNIV. OF
CALIFORNIA

PLATE IX.

APPLE MOTH---CODLING MOTH.

COMMONLY CALLED APPLE WORM.

PLATE IX.

1. The Larva or Caterpillar of the Apple Moth resting on a part of the Core of an Apple.
2. Pear, July 6, showing the Borings of the Apple Worm.
3. Pear, July 18. The dark spots near the blossom end indicate the worm at work under them. There is usually a depression at this part.
4 and 5. A very common appearance of the early Summer Apples from the operations of the Apple Worm during the month of July. The Worm, Fig. 1, was taken from the Apple, Fig. 4, and is not quite full grown.
6. Is the half of a Westfield Seek-no-further, cut open March 17, and is an exact representation of the original. This shows that even such injuries will not always prevent the fruit from keeping through the winter.
7. The Moth at rest, natural size.
8. The Moth with the wings expanded. These were painted from living specimens on the 18th of June.

MORRIS calls this moth *Penthina pomonella* (*Carpocapsa*). Emmons, *Carpocapsa pomonella*. Say does not mention it in his works. Harris says, " it is not a grub but a true caterpillar, belonging to the *Tortrix* tribe, and in due time is changed to a moth called *Carpocapsa pomonella*, the Codling Moth, or Fruit Moth of the Apple." Kirby and Spence speak of it as one of the enemies of the Apple in England. Reaumur gives its history, and says, " It is a species of moth common in Europe (*Carpocapsa pomonella*), the caterpillar of which feeds in the centre of our apples, thus occasioning them to fall." An anonymous writer in the *Entomological Magazine* of London has well remarked, " that this Moth is the most beautiful of the beautiful tribe to which it belongs; yet, from its habits not being known, it is seldom seen in the moth state; and the apple-grower knows no more than the man in the moon to what cause he is indebted for the basketfuls of worm-eaten windfalls in the stillest weather."

In a very careful examination of large collections of pears and apples sent from France to the exhibitions of the American Institute, I have found the marks of the Apple Moth in some, but in no instance have I seen the crescent mark of the Curculio.

As the result of experience, founded upon close observations for a number of years, and extended through large sections of several states during the summer and fall of 1864, I have come to the conclusion that this Apple Worm, as it is generally called, is as destructive to apples, pears, and quinces as the Curculio, but not so to the stone fruits. When we shall have subdued the latter we shall have cherries, plums, peaches, apricots, and nectarines, without much further trouble from insect enemies; but we must control both the Curculio and Apple Moth before we can secure the apples, pears, and quinces.

Although the Apple Moth is an imported insect, it seems to have become as widely extended as the native Curculio. When we see a butterfly fluttering about our fields or gardens, and know that it lives a very few days, we would think that any one species, starting from a given point, would be very slow in spreading over a continent. What is the rate of speed of a butterfly, or the length of time it can continue its flight, I have never seen estimated. Most kinds are visible for a moment, and then out of sight. Probably no calculation has ever been made that would approximate the truth. Kirby speaks of the male of one of the Silk Moths (*Attacus paphia*), as supposed to be capable of a flight of one hundred miles. A dragon-fly will hover for hours on the wing over a pond of water, passing rapidly to and fro in pursuit of insects. One species of this tribe has been known to alight on ships 500 miles from the nearest shore. But insects of this order greatly excel the butterflies in wing power; they have the speed of the fleetest birds, and more than their quickness in turning. If we know little of the travelling power of the butterfly whose flight is by daylight, how much more difficult will it be to observe the flight of moths, which only takes place at night.

The caterpillars of both butterflies and moths are nearly all vegetable feeders. Many species feed on one kind of plants, as the silk-worm on the mulberry; some on two or three, as the tent caterpillar, which will grow to maturity on the apple and wild cherry, but will starve on the pear. Other caterpillars will feed indiscriminately on the leaves of many trees or plants, provided these leaves come early enough.

The larva of the Apple Moth, like the grub of the Curculio, has usually a fruit to itself, the parent moth depositing but one egg on a fruit, and it is supposed that another Apple Moth will seldom duplicate that egg.

The first part of the life of this caterpillar is usually passed in feeding on the substance of the fruit near the blossom end, and while there it is quite small. Afterwards it will be found in and around the core. The holes drilled through the pulp are tunnels for passage only, not excavations made in feeding—the contents being a mere pomace, and not the castings of the insect. This indicates that the chief food this caterpillar requires is to be found in the core, including the seeds, and is in limited supply; hence we seldom meet more than one in each fruit. If the whole pulp of the fruit were suitable for food, most of our apples and pears would afford ample nourishment for a dozen of these worms.

The Apple Moth, like most other moths and butterflies, has a great number of eggs to dispose of. She will have the appropriate nidus for her young if she can find it; and how far she will go in pursuit of Apples, Pears, or Quinces, if there should be none near her native tree, or if they have all been appropriated by others, before she was ready, is a difficult question to decide. One of the most interesting subjects of contemplation to the naturalist is to watch the movements of moths in the dusk of summer evenings. They will slow up to a plant or tree, as a steamboat to a landing—merely touching, then on again to another, and again and again, till they find what they want, deciding, as they go, whether the leaves that come upon those trees after an intervening winter will be the proper food, or will appear early enough for the little ones that are to issue from their eggs.

Kirby and Spence say that the progress of the Hessian Fly was at the rate of fifteen or twenty miles a year. Dr. Fitch, in his most valuable account of the Wheat Midge, says, that the spread of this insect along the country bordering the St. Lawrence and Lake Ontario, was at the rate of about nine miles a year. But the history of the appearance of these two insects, like that of the Apple Moth, in the different parts of the country which they have visited, shows that they had no fixed rate of progress. Speculations as to where an insect came from, or when it arrived, or at what rate it can travel, will avail but little as to this Apple Moth pest. *It is here, it is all over our country*, wherever Apples and Pears are cultivated, in many places appropriating half these crops every year, and it is rapidly increasing. While the two wheat enemies have been subdued in a great measure by parasites, nothing of the kind has made much impression on the Apple Moth; and from its habits of life we have little reason to hope for relief in that direction. We must help ourselves, and the sooner we begin the better.

Figure 1, Plate 9, represents the larva or caterpillar of the Apple Moth resting on a portion of the core of an apple. It is usually five-eighths, sometimes three-quarters of an inch in length, and is about one-eighth in thickness or diameter—nearly double the size of the grub of the Curculio. It is of a reddish color, often a decided pink. This worm has all the characteristics of a caterpillar; six true legs at the head end of the body, and eight prop or fleshy legs. The head is sometimes dark-brown, and sometimes glossy black. It is to some extent a silk-making caterpillar. Throw it off suddenly from its resting-place, and it will often let itself down with a cord, as a span-worm does. This will never be done by the grub of a beetle or the maggot of a fly. It will not go into the ground, as the grub of the Curculio does, but will climb up the body of a neighboring tree. The Pears, Figures 2 and 3 of this Plate, are the Beurré Clairgeau; a variety that often suffers from the depredations of this enemy, though not as often as the Bartlett, or some of the small early kinds. The borings on Figure 2 are an uncommon appearance, and show where a well-grown worm has made an entrance from the outside. The borings of the Apple Worm are nearly always found in the shape of a round plug of the pulp of the fruit, forced out from within through a hole about the size of one made by a small gimlet. This indicates that the worm or caterpillar has come to its growth, and will soon push out this plug entirely, and then escape. Whether this worm leaves the inside of the fruit at night only, has been a subject of investigation to some extent, but I have not evidence enough to establish the point. It may be an instinct in common with many other insects, to leave such retreats only in the dark, when it will be more secure from bird enemies. I certainly often see these plugs filling up the holes in the daytime, and find next morning that they have been pushed out, and I have never seen one of the worms escaping during the day.

The appearance in Figure 3 is a very common one. The Moth usually deposits her egg at the blossom end of the young fruit; and just within the calyx is a tender spot, where the minute larva from that egg finds an easy entrance to the interior. In that part of the fruit under these dark spots it will be found feeding until it is a quarter or one-third grown, making quite an excavation. This part ceases to grow or expand, and there will soon be a depression. The spots indicate that the worm has approached to the skin itself as far as the black extends. The young caterpillar can be taken out without making much of a wound, but this operation seldom saves the fruit. Soon after this it will be found in the centre or core, making extensive excavations, involving the seeds, as seen in Figures 4 and 6. This little miner often shows much ingenuity in keeping its apartment in order. If its chips and castings were permitted to lie about loose, they would be inconvenient, and in windy weather annoying. To guard against this, it ties them all up together with silken cords, and then secures the mass to a part of the establishment most out of the way.

Figures 4, 5, and 6, do not by any means indicate all the forms of injury or deformity caused by this enemy. Twenty illustrations could be made, all differing in appearance, but each having some characteristic, proving it to be the work of the larva of the Apple Moth. The seeds of both apples and pears will often be found to have been eaten as seen in Figure 6, on this Plate.

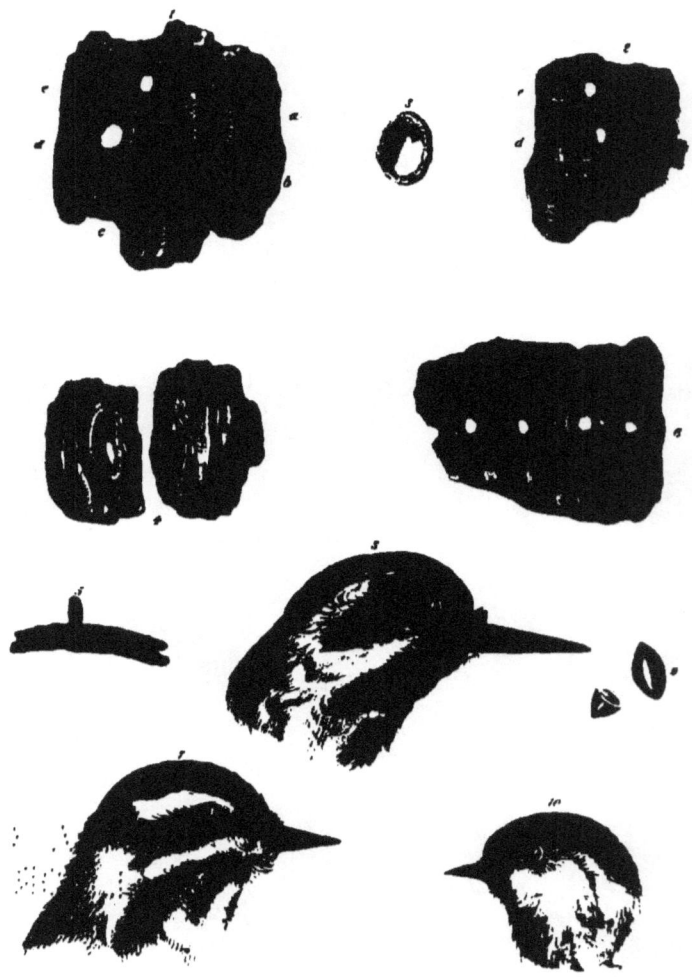

PLATE X.

1. Is the under side of a scale of bark taken from an Apple tree.
    - *a.* Larva of Apple Moth exposed by a cocoon being opened when separating the bark from the tree.
    - *b.* Represents a cocoon entire, and the larva within.
    - *c.* Shows a portion of an old cocoon.
    - *d* and *e.* Show where the cocoons have been penetrated, and the larvæ taken out by the Bird—Fig. 7.
2. Shows the outside of a part of the scale, Fig. 1. *d* and *e* are holes corresponding to *d* and *e* in Fig. 1.
3. Represents a ring formed from the pellets or chips taken out of the cavities, and tied together with silken cords. This ring usually fills up the space between the scale and true bark, making the cocoon perfectly tight, and to some extent binds the scale to the tree.
4. The pupa of the Apple Moth.
5. The pupa case after the Moth has escaped.
6. Holes made in parallel rows by the Bird—Fig. 8. This kind of holes gave rise to the name, "Sapsucker," applied to the bird supposed to make them.
9. One pit and the half of another found in the stomach of the Bird—Fig. 8.
10. The head of the Black Cap Titmouse, or Chick-a-dee.

This Plate is a study in itself, and a careful consideration of it will show the intimate connexion between Ornithology and Entomology.

Fig. 1 is a representation of the inside of a scale of bark from an Apple tree, and Fig. 2 is part of the outside of the same scale.

One who is curious to ascertain the winter homes of insects will find many species secreted under the scales of the bark of trees. In separating these scales of Apple or Pear trees in search of the larvæ of the Apple Moth, he will find lady-bugs, spiders, flies, collections of the eggs of spiders nicely arranged together in nets made of the finest materials, many kinds of small beetles, &c. That a large number of the Apple worms conceal themselves under the scales of the bark of the trees in an Apple or Pear orchard is certain; but where they go when the trees have none of this rough bark is not so clear. They do not form their cocoons upon the ground or in the grass. Though caterpillars, they are totally without the coating of hairs that prevents so many other species from becoming the food of birds, and would be most tempting morsels for both birds and poultry. Ants also would be formidable enemies of such worms. To ascertain how their instinct of self-preservation would manifest itself, I have often collected a number, and placed them in the vicinity of a tree. They creep about at random for a little while; but if not too far off, most of them will soon be seen going in the direction of such tree.

In the spring of 1860, when the seventeen-year locusts were coming up out of the ground, I often tried this experiment with them, and uniformly with the same result. Put them down anywhere within ten feet of a tree, their course would soon be directed towards it; and no matter how often they fell back in struggling over the grass and other obstructions, nothing diverted them from their path. Whether they could see or not is hard to say. Until within a few minutes they had never been in daylight; eyes to them, in all their seventeen years' experience under ground, would have been as useless as to the fishes in the Mammoth Cave. How long the apple-worm caterpillar will creep about—how many trees it will ascend and descend in search of this place of concealment, I do not know; but this instinct would indicate a perseverance till the end was accomplished. If we had no way of trapping this enemy it would teach us to keep our trees clear of all rough bark, let the poultry have free access to the orchards, and protect the birds. *a*, Fig. 1, is the caterpillar or larva of the Apple Moth as seen in one half of its cocoon. It has been thus exposed in taking off this scale from the tree. Fig. 3 is a part of its cocoon. The ring is formed by the chips or nibblings that it makes in digging out the little cavities; and this ring, or ridge, fills up the space between the scale and true bark. Both saucer-shaped cavities are usually lined with a delicate silken cocoon. The part of this cocoon attached to the ring is shown in this figure. (It must be remembered that this ring and the part of the cocoon attached, made the other part of the house that covered snugly this now exposed worm.) These Apple Moth larvæ, such as *a*, Fig. 1, can be found in this stage at least nine months of the year. Those that come out of the later apples and pears remain as caterpillars through the fall and winter, and many of them till quite late in the spring; then, during two or three weeks, they will be found in the pupa or chrysalis state, as shown at Fig. 4. Then in June the cases will often be seen, as shown at Fig. 5, the moth having escaped. In taking off scales of bark at this time, or even in approaching a tree, the moth will often flutter away almost unperceived. It is so nearly the color of the bark as seldom to be noticed when at rest. This is characteristic of many moths; though not Chameleon-like, or having the power to adapt themselves in color to the object on which they rest at the time, they are so strikingly like the bark of the trees on which we find them as nearly always to escape our notice. This insect in its pupa state, like many others, will be found torpid and apparently helpless in its cocoon; but when its time comes to escape from this mummy condition it has some locomotive power, and by a kind of wriggling motion forces itself to an outside opening. This is a necessity to the perfecting of the wings; they could not be expanded in so confined a space. The silk-worm moth liberates herself from her cocoon by burning a passage-way with an acid. The dragon-fly, in its immature condition, is a kind of bug and lives in water; when about to become a winged insect it will creep up a rush or reed until out of water, and make itself fast to this reed by its claws, when the back of its sub-marine case will open, and the perfect insect emerges. That is what the seventeen-year locust comes out of the ground for. The young mosquito floats to the surface of the water, and the sub-marine coat splits open. First her head, then her body emerges; finally, she will be seen standing straight up in her former skin like a mast in a canoe. All these insects—and many more could be mentioned—assume this perpendicular position immediately on leaving their pupa cases. If watched, the little wings that had hitherto been compressed into the smallest possible space will now be seen to enlarge gradually; there will be an occasional slight flutter. The fluids of the body would seem to be settling there—*gravity aids them*—and soon they will be expanded to their utmost limits. Such wings as those of the Apple Moth could never have been unfolded in the cocoon between these scales of bark. She does not resort to the appliances of chemistry, as the silk-worm does in the use of acids; but she is equally philosophical, and appeals to the force of gravity.

The life of this insect as a moth is short. If she came to this last and perfect condition early in the season, the fruits would not be ready for her, and she would die before fulfilling her mission. Nature makes no such mistakes. The fruits will be ready for the moths when the moths are ready for the fruits. In a few weeks the caterpillar or worm from the egg of this moth will be matured, and found in its cocoon under a scale of bark, appearing as a moth in August; living as a caterpillar not half as many weeks as the winter caterpillar did months; the one living in a temperature far below zero, and frozen solid as ice; the other knowing the hottest weather of our climate. That "the wind is tempered to the shorn lamb," is an expression often used, and by many it is believed to be scriptural. The winds pay very little regard to insects, but insects are wonderfully adapted to the winds. *b*, Figure 1, is a representation of another cocoon; one that has not been injured in taking off the scale of bark, and the larva within not disturbed. This cocoon is of an unusual shape, but the builder was evidently crowded by neighbors, and had to be guided by circumstances. *c* shows where one has been that has escaped as a moth. *d* and *e* in Figure 1, and *d* and *e* in Figure 2, indicate holes in the inside and outside of the same scale of bark. These holes have been made directly into the cocoons of these caterpillars, and those cocoons robbed of their contents by the bird, Figure 7, the Downy Woodpecker. The holes in the piece of bark, Figure 6, are different. They are parallel, and have not been made in a dry scale, but in the green bark, where no insects live. Figure 8 is the bird that makes these—the Yellow-bellied Woodpecker. Figure 10 is the Chick-a-dee, an occasional feeder on the Apple worm. These heads, as represented in the colored edition of this work, strikingly resemble the originals; but the ornithologist will notice defects. They do not look as the same birds do in Wilson or Audubon. The positions are not similar. The plumage also differs, but that may be owing to the age of the bird. An ornithologist chooses his specimens from birds at full maturity, and at a season of the year when the plumage is the most perfect. These were shot to ascertain what their stomachs contained; and as two of them were proved to have been usefully employed, their likenesses were painted to commemorate their good deeds, and not to exhibit them as types of beauty. Our investigations in regard to the insect enemies of fruits would be only partial, if no attempts were made to ascertain how far the birds are useful to us in controlling them.

We can approach some of the smaller birds so closely as to be able to see what they eat. A field glass will often aid us in such investigations; but this source of knowledge is not always reliable. The works on ornithology give us some information on the food question. Wilson, Audubon, and others, often accompany their figures of birds with illustrations of fruits, berries, and insects, intended to show that these are the food of such birds. But this subject is not a primary consideration in any of these works. The Ninth Annual Report of the Massachusetts Board of Agriculture contains an admirable article of about thirty pages on the usefulness of birds, by Wilson Flagg. Of the many contributions to the history of birds, I have met with none so interesting as this. Some French investigators have made elaborate reports within a few years on this subject; but the birds of Europe are so different from ours that such reports can give us little practical information, unless accompanied with illustrations, and especially showing the form of the beak of each species.

To make such a work as this complete, requires more exact and positive knowledge than could be procured from any of these sources; and I have killed a very large number of birds and examined the contents of their stomachs, especially of those frequenting orchards. Most of these examinations have been made with a magnifying glass, and many with the microscope. Some species I have shot at short intervals during the season, to know how far their food varied at different times; and I have thus ascertained that the contents of the stomach at any one time are not an infallible criterion by which we can determine the usual food of that bird. On the fifth of May, 1864, I shot seven different birds; they had all been feeding freely on small beetles, and some of them on nothing else. There was a great flight of these small beetles that day; the atmosphere was teeming with them. A few days after the air was filled with ephemera flies, and the same species of birds were then feeding upon these.

The killing of so many birds has been a most repugnant task; but I have nerved myself to it in the cause of science. I felt there was a want of such information, and once procured it could not be wanted again. The comparatively few thus sacrificed would become martyrs for the good of the many. Many of these investigations have been of surpassing interest, from the consciousness that such knowledge, if properly disseminated, would create a public sentiment even stronger than law, for the protection of the birds.

I have found in the Baltimore Oriole the remains of Curculios, the real plum weevils. The Downy Woodpecker and the Chick-a-dee eat the caterpillar of the Apple Moth. The Oriole, Wren, and Cat-bird know how to find the leaf-curling caterpillars in their places of concealment, and thus protect our orchard and garden trees and shrubbery from much deformity. The Cedar birds come in flocks for the span-worms. Even the Bob-o'-link does not perch himself on our apple trees exclusively to show off his fine feathers or charm us with his unpronounceable music, but makes a meal occasionally upon the canker worms. The beak of this bird is much like that of the sparrow or canary, and formed for husking seeds; still I have found his stomach filled to repletion with these troublesome caterpillars.

The season of 1864 will be memorable as the year of Aphides, or plant lice. The first crop of leaves on many of the Apple trees was so alive with a species of these pests that most of them fell off, causing also a profuse shedding of the young apples. Warblers of many kinds, then just coming on from the south, Creepers, Wrens, and even Sparrows, as well as many other kinds of birds, fed upon these the livelong day. The throats, and even the back parts of the beaks of some of them, would be found lined with these aphides, many of them still alive, and their stomachs containing a juice that would leave the hands colored as they are after crushing these insects. The creases or folds of the stomachs were lined with what appeared to be an accumulation of the hairs of caterpillars, but under the microscope were found to be the legs of these plant lice—thousands and thousands of them.

The account of almost every insect that will pass under review in the progress of this work, will contain a monograph of some bird that has been found its special enemy. Some will be shown of full size, and in positions exceedingly graceful, illustrating surprising intelligence as to the manner of finding and securing their insect food.

EXTRACTS FROM DIARY OF 1864.

"PICUS PUBESCENS—*Downy Woodpecker; Sapsucker.*—Sp. Ch.—A miniature of *P. villosus.*—Above black, with a white band down the back. Two white stripes on the side of the head; the lower of opposite sides always separated; the upper sometimes confluent on the nape. Two stripes of black on the side of the head, the lower not running into the forehead. Beneath white; wing much spotted with white; the larger coverts with two series each; tertiaries or inner secondaries all banded with white. Two outer tail feathers white, with two bands of black at end; third white at tip and externally. Length about 6¾ inches; wing 3⅞. Male with red, terminating the white feathers on the nape.

"*Hab.*—Eastern United States, towards the eastern slope of the Rocky Mountains."

The above is Baird's description of the Downy Woodpecker (the head of which is shown in **Fig. 7** of this Plate), and has been selected for this work by a friend who is a most accomplished ornithologist.

The bird that knows how to find this formidable enemy of the fruit-grower, and destroys it in such numbers, is an object of special interest, and a scientific account in a work like this becomes a necessity, that it may be positively identified.

DOWNY WOODPECKER.—*April* 8.—In shooting a Robin to-day, on an Apple tree, I started one of these little birds. It flew to the next tree, and I secured it with the other barrel. This was the first of this species of bird I had been able to approach in the orchards near the city. The stomach contained *one larva of an Apple Moth* so nearly entire as to be easily identified. The other contents of the stomach were the remains of small beetles, some of which had those brilliant metallic hues so difficult to describe and so impossible to imitate by paint. The head of one of these, under the microscope, equalled in lustre the diamond beetle. No seeds or sand, and no signs of sap or the sap bark of the Apple tree could be discovered.

April 21.—In a ramble to-day on the borders of the Passaic River, some ten or twelve miles from Newark, where the birds are more plentiful and not so wild as nearer the cities, I shot another of these Downy Woodpeckers. The stomach contained one beetle, the *heads of the larvæ of two Apple Moths*, and the heads of the grubs of three small Borers. The bird, when first seen, was pecking or sounding an old fence post, and then flew to an Apple tree. There was an orchard of some twenty or thirty old Apple trees here.

May 3.—Shot another Downy Woodpecker to-day. It had been eating several black beetles, and three grubs, but they were not the larvæ of the Apple Moth.

Aug. 5.—CROOKED LAKE, *Yates County, New York.*—The bird that knows how to find the Apple Worm under the scales of bark on the trees, has been here. I find the unmistakable mark—the round hole in the scale leading directly into the place where the worm had been in its cocoon. The parallel lines of holes are on the Apple trees here also.

Nov. 6.—In a ramble in the country to-day I saw a couple of Downy Woodpeckers. I watched them for some time, but could see nothing in their actions to indicate how they found the Apple Worms under the scales of bark.

Nov. 10, 11, and 12.—During an excursion in the upper part of Morris County,

N. J., made for the purpose of investigating the insects and birds, I had an opportunity of seeing a Downy Woodpecker in an old orchard, and passed an hour watching his actions. His creeping power is wonderful. (His foot should have been represented in this Plate, to show how perfectly it is formed for the purpose.) I was especially interested to see with what speed he could move down the body of a tree backwards. This seemed even more rapid than the forward motion.

Here I was gratified in being able to ascertain how he finds where to peck through the scales of bark, so as to be sure to hit the Apple Worm that is so snugly concealed beneath. The sense of smell will not account for it. Such an acuteness of one of the senses would be beyond the imagination. Instinct, that incomprehensible something, might be called in to explain to those who are satisfied to have wonders accounted for by means that are in fact only confessions of ignorance. Birds have instincts undoubtedly—so have we; but they are mixed up confusedly with other faculties. Most of the actions of insects are purely instinctive and utterly unaccountable. But the Apple Moth is not a native of this country—the Downy Woodpecker is. The bird would not have been created with a special instinct to find the larva of a moth that did not exist in the same country. Other insects live under these scales of rough bark; but in very numerous examinations, I have not seen such a hole made except when leading directly into the cocoon of this particular caterpillar.

This little bird finds the concealed larvæ under the bark, not from any noise the insect makes; it is not a grub of a beetle having a boring habit, and liable to make a sound that might betray its retreat, in seasons of the year when not torpid. A caterpillar makes scarcely an appreciable noise, even when spinning its cocoon, and when that is finished it rests as quietly within as an Egyptian mummy in its sarcophagus.

There is no evidence that the Downy Woodpecker ever makes a mistake; it has some way of judging. The Squirrel does not waste its time in cracking an empty nut. There is no reason to believe that this bird ever makes holes through these scales merely for pastime, or for any other purpose except for food. He knows before he begins that if he works through, just in that spot, he will find a dainty morsel at the bottom of it, as delicious to him as the meat of the nut is to the squirrel. But how does he know? *By sounding*—tap, tap, tap, just as the physician learns the condition of the lungs of his patient by what he calls percussion. The bird uses his beak, generally three times in quick succession—sometimes oftener; then tries another. Watch him. See how ever and anon he will stop in his quick motions up and down, and give a few taps upon the suspected scale, and then test another, and another, until the right sound is communicated to that wonderful ear.

Here is evidence enough of the usefulness of this bird to entitle it to exemption papers for ever. Reader, look carefully at the head, as represented in Fig. 7 of this Plate. Do not call that bird "Sapsucker." That name will create a prejudice with some. The whole tribe of Woodpeckers labor under a prejudice in some neighborhoods. Some will eat cherries, and some are supposed to be fond of grapes. But the chief food of all of them is insects, and many of those insects are our worst enemies. It will be well to let all the Woodpeckers have their own way, but by all means protect the Downy.

Fig. 8, is the SPHYRAFRICUS VARIUS of Baird; *Picus varius* of Wilson, Cassin, and Audubon.

YELLOW-BELLIED WOODPECKER.—Sp. Ch.—Fourth quill, longest; third, a little shorter; first, considerably shorter. General color above, black, much variegated with white. Feathers of the back and rump brownish-white, spotted with black. Crown, scarlet, bordered by black on the sides of the head and nape. A streak from above the eye, and another from the bristle of the bill, passing below the eye and into the yellowish of the belly, and a stripe along the edges of the wing coverts, white. A triangular broad patch of scarlet on the chin, bordered on each side by black stripes from the lower mandibles, which meet behind and extend into a large quadrate spot on the breast; rest of under part yellowish-white, streaked on the sides with black. Inner web of tail feather white, spotted with black. Outer feathers black, edged and spotted with white. Length, 8.25 inches : wing about 4.75; tail, 330. Female with the red of the throat replaced by white. Young male without black on the breast, or red on the top of the head.

Hab.—Atlantic Ocean to the eastern slopes of the Rocky Mountains; Greenland.

Oct. 6.—Went to the orchard of a friend in the outskirts of the city, and found a Sapsucker (or, as it should be called, a Yellow-bellied Woodpecker) at work on one of the Apple trees. It had made almost three hundred marks (see Plate 10, Fig. 6). Some were new and others old ones drilled out. This was in the forenoon. I watched it about half an hour. I returned in the afternoon. The same kind of bird was on the same Apple tree, and busily at work pecking holes. I watched for about an hour, sometimes approaching so near that it would fly a short distance and watch me closely till I would go thirty or forty yards from it, when it would at once return and resume the pecking work. If I were in sight it would keep on the opposite side of the limb, occasionally peeping round cautiously to see if I was coming nearer. Sometimes, during the intervals of these peeps, I would quietly approach so closely that the moment it saw me it would fly away, but to return again as soon as I had retired to a proper distance. After bo-peeping till I had no more time to spare, I shot this poor bird, expecting to find positive evidence in the stomach of what it made these holes for—and found two seeds or pits (of which one and half the other are represented by Fig. 9, Plate 10), with the purple skins of the same fruit, seven small ants, and one insect of the chinch bug kind about the size of those found in the beds of some taverns. But of bark or sap there was not even a trace.

Later in the day I shot another of the same species of bird in an old orchard out of town. The stomach of this one contained the pulp of an apple and one ant—nothing else. This one was on the upper part of an Apple tree, and was not pecking or sounding. The investigation of this bird so far is unsatisfactory. I have seen no evidence yet that these holes are made in search of food. Ants are certainly found sometimes about these holes, and apparently in pursuit of the sap that exudes from them; but the idea suggested by some, that the birds make them to attract these ants by such tempting baits, is a palpable exaggeration of the reasoning power of this bird.

From what I have seen to-day, as well as from former observations, and from the testimony of several careful observers among farmers of my acquaintance, I am led to believe that Baird is mistaken in calling the preceding bird—the Downy Woodpecker—a Sapsucker. Wilson evidently believes the same thing, although he does not say so in express words. The bird that makes these parallel holes, as shown at Fig. 6, Plate 10, has a bad reputation, and I am anxious to relieve my little friend that finds the Apple Worm, from all charges that will bring him into trouble. I believe that this bird, Fig. 8, Plate 10, is the only one that makes such holes. The Downy Woodpecker, Fig. 7, Plate 10, makes many holes in Apple and Pear trees also, but not in this regular manner. The one is in search of the larvæ of the Apple Moth under the dry scales of the bark; the other seeks—I don't know what, in the green bark itself.

These parallel rows are often very numerous. The trunks and larger branches will often be seen covered with them. The rows will be almost adjoining for many feet, and the holes in the rows as near each other as represented in Fig. 6 in this Plate, and running all round the tree. In old orchards where trees have been grafted high up, these holes may be sometimes seen in the stock and not in the graft, and sometimes in the graft and not in the stock. Rows of trees of one kind of fruit will be filled with these holes, while adjoining rows of other kinds will be exempt. So far as I have noticed, these holes have been made only in October. Sometimes they will be seen in Cherry trees, and I have observed those trees thoroughly drilled. Some Evergreens are so pecked in this way as to bleed the next season to an injurious extent. But I have not been able to ascertain that they impair either the growth or the fruit-bearing power of the Apple tree.

The grub of the Apple Tree Borer works between the bark and the wood during the first of the three years of its life, but it is always either under or so near the surface of the ground as not to be likely to be found by the Woodpeckers. I have never seen any bird in pursuit of this grub. The two following years it is always so far within the wood as to be out of reach of the birds. I know no other grub that works under the bark of living Apple trees; but an Apple tree in decay will often be found teeming with grubs.

Figure 10, PARUS ATRICAPILLUS—Linnæus. *Black-cap Titmouse*—Wilson. *Chick-a-dee*. Sp. Ch.—Second quill long as the secondaries. Tail very slightly rounded; lateral feathers about ten, shorter than middle. Back, brownish ashy. Top of head and throat black, sides of head between them, white. Beneath, whitish; brownish white on the sides. Outer tail-feathers, some primaries, and secondaries, conspicuously margined with white. Length, 5; wing, 2.50; tail, 2.50. *Hab.*—Eastern North America along the Atlantic border.

March 7, 1864.—The Chick-a-dees are flitting about upon the Apple trees in the orchards and Maple trees in the swamps—chick-a-deeing and very happy, and why not?—in these bright warm days after the long cold winter. This is one of the creepers—quick in all its motions.

March 8.—Shot a Chick-a-dee to-day, but the contents of its stomach were so comminuted that it was impossible, without the aid of the microscope, to distinguish anything positively. Some portions looked like parts of beetles, and others seemed like the broken shells of the eggs of moths and butterflies. Could see no seeds.

The habits of these minute little friends are delightful. You see two, three, four, sometimes more—in an Apple tree, climbing up pendent twigs and examining them all round, apparently in search of the eggs or minute larvæ of insects. Next you will see their little beaks working among the moss on the older branches. Then they will be pecking at something like a beetle in a crotch. Next there will be a gentle whistling call, and all will be off to the next tree. Often other birds, especially sparrows, will be seen closely following these restless little fellows, as if attracted by their chick-a-dee melody. These are winter birds; and though differing from the sparrows in being insectivorous, in other respects they are so much alike as to be fond of each other's company—though it appeared to me that the sparrows were more partial to the chick-a-dees than the chick-a-dees to the sparrows.

March 28.—In a ramble through the orchards to-day I saw but one of these birds, and this one passed so rapidly from one tree to another, there was but little opportunity of watching its habits or seeing what kind of food it was in pursuit of. Its song has changed since last visit. Then it was chick-a-dee; now it is chick-a-dee-dee-dee-dee-dee, and sometimes dee-dee-dee-dee-dee-dee.

April 1.—Shot one to-day, and, what is of great importance, found *five of the larvæ* of the *Apple Moth*. One of these had been so recently taken, and was so little mutilated, that it was easily identified. The heads of the other four appeared identical when examined with a pocket-glass; but when subjected to the test of the microscope, there was no possible room to doubt. The day has been dry and windy, following a warm wet day and night; and it is in just such weather that the bark of the Buttonwood, Shell-bark Hickory, and other shaggy trees, will be found curling out and falling off.

I have never seen anything that would lead me to believe that this minute bird makes the holes in the scales of bark that lead directly to the cocoons of these caterpillars; they are made by the Downy Woodpecker, and probably by it alone. The Chick-a-dee most likely finds these worms only or chiefly on such days as this, when the warping of these scales exposes them to the prying eyes of these busy little friends. This bird is one of the guardians of the orchard; quick, active, always on the alert; assuming any position; sometimes even hanging by one foot on the under side of the large limbs, where these caterpillars rather prefer to conceal themselves; and now proved to feed freely upon the second in importance of the insect enemies of our fruits. Let no one hereafter kill a Chick-a-dee without being made to feel that he has done a most disgraceful deed.

Nov. 6.—Have been making a short trip to the central part of the State. See no birds now except crows, doves, quails, sparrows, and the creepers, including the Chick-a-dees. The last are very numerous—almost in flocks. They are quite tame. I have been much amused at a little party of them on the long, slender shoots of some swamp-willow sprouts. One of them would alight on the extremity, and then the twig would bend until the bird looked as if it were holding on to the end of a suspended string; but it was perfectly at home, hardly even fluttering to maintain its position, and there it remained pecking away among the buds at the extreme end of the shoot, apparently in pursuit of plant lice so often found in just such places. We are told that the "tender mercies of the wicked are cruel;" my tender mercies overcame the cruel on this occasion; and although I wanted to know exactly what the bird could find there to eat, the gun would not point in that direction.

Nov. 15.—I have never before noticed the Chick-a-dees so numerous as this fall. Shot one to-day; it had eaten four small seeds, almost as hard as gravel stones, and quite a number of the pupæ of very small beetles, such as take shelter under moss and old bark on trees.

Jan. 1, 1865.—For a month past the Chick-a-dees could be seen in the mornings on the Elm trees of this city—always on the slender twigs, and busily searching round the buds. The weather seems to make no difference. The gentle, plaintive call by which they keep in company could be heard every morning, even in the most pitiless storms. In the evenings they could be seen congregating about the evergreen trees.

For several mornings in succession I noticed that the piazza was strewn with the cocoons and broken pupa cases of the caterpillars that were so numerous in September; sweep them off, and soon they would be there again. It was the work of the Chick-a-dees. The piazza is a high one, and extends on three sides of the house. Hundreds of caterpillars formed their cocoons in the chinks and crevices of the ceiling, and there these little birds found them. I hung out pieces of fat pork, and bread and butter, and they tasted moderately; but as soon as the pupæ of these caterpillars were all consumed, my kind of food was neglected.

March 1.—I have seen a small party of the Chick-a-dees on the Elm trees again to-day, the first for several weeks. They were examining the buds.

PLATE XI.

PLATE XI.

1. Represents the hay-rope trap, slipped up a few inches above where it had been set.
2. Shows where this hay-rope band has been during the season. The marks are intended to represent the slight concavities made by the Apple Worms under the rope.

This Plate represents a Tree of Fall Apples, as it appears where the Curculio and Apple Moth are not interfered with, but are permitted to have things their own way.

This tree is intended to represent the one so often spoken of in the following diary, on which were caught nearly two hundred of these Apple Worms during the season of 1864.

These bands should be put on the trees as soon as the fruit shows signs of the worms being at work, as seen in the illustrations in Plate 9;—from the middle to the last of June. They should be examined every two weeks, as long as warm weather lasts, the earlier broods of worms becoming moths, and producing a second crop. If the orchard is pastured the bands must of course be put out of reach of the animals. Sometimes it may be necessary to place them round the limbs; in that case the scales of bark on the bodies of trees below them should be scraped off.

Those who watch the falling of the blossoms from fruit trees will notice that the greater number leave no embryo fruits. A few days later a large portion of the young fruits that had formed, cease to grow, and they soon fall off. These are blights, and this is nature's mode of relieving the parent of the burden of too large a family. Occasionally there will be a frost just as the petals of the blossoms have fallen off, and the germ being exposed at its most tender age, sometimes the whole crop will perish.

In the spring of 1864, when the Pear, Peach, Plum, and Cherry trees were just shedding their blossoms, there came several days of very warm wet weather, with scarcely any wind; and during this time a dense fog prevailed for many hours. The calyx surrounding the embryo fruit was now like a sponge, absorbing the rain, and there being no chance to dry during all this time, the fruit germ rotted. By separating the wet calyx from the young fruit it was easy to see the commencement of this decay; it was generally in streaks. On the Cherry trees the effect was singular. Wherever the young cherries came in contact with the leaves the decay was communicated to them, and they were left with holes, presenting a ragged appearance. Half of the above kinds of fruits perished in that way; and the trees of some kinds of cherries, where the bloom had been profuse, had absolutely no fruit left.

The season of 1864 will be as memorable for the plague of plant lice on our fruit trees as that of two or three years before had been for a visitation of a similar insect on the wheat and oats; and excepting as they were fed upon by small birds and some insect enemies, there seemed to be no hope of saving the Apple crop from these minute enemies.

The Peach crop is often destroyed by the buds being killed by severe cold in winter, and sometimes by the young leaves being so diseased by a "curl" that the fruit will nearly all fall off when quite small. For all these accidents the fruit-grower should not be held to too rigid an account, though some of them might be guarded against by a judicious choice of a situation for the orchard. But the fruit-grower who lets his Peach and Apple trees be girdled by Borers; who permits his orchards to be overrun by the Tent Caterpillar, and his Plum and Cherry trees, to become masses of knots, or his young trees generally to be sapped by millions of bark lice, deserves little commiseration.

The tree in the preceding Plate had escaped all these contingencies, and showed, until some time in June, a promise of a most bountiful crop; but then the young apples began to fall, and persevered in falling till not a dozen were left to come to full maturity.

Now let us imagine the owner of an orchard, who has taken the best possible care of his trees for ten or fifteen years, finding every season the entire crop lying on the ground when not half grown, and of no value—and the imagination need not go far to find such an owner—Does he feel comfortable? Perhaps his annual expenses overrun his income, but a fair crop of fruit would have reversed his financial condition. It was not the fault of the orchard; the trees were full enough at one time. Why did they all fall off?

I have read somewhere that there was once a man who owned a cow—

"But had nothing in the world to give her;
And he said, consider, cow, consider."

It may have been the sight of the hay-band in this Plate that brought this old-cow poetry to my recollection; or it may have been the wormy apples under this tree that I am so anxious the cows should have a chance to eat. The word "consider" is very appropriate here. And now let us go to work and "consider" this matter very seriously.

A young man is just starting in life. He has bought a farm that is to be the permanent home of himself and wife. He has probably gone in debt. The parents on both sides have contributed furniture, and horses and cows, and farming utensils. A comfortable house has been built. But there is no orchard—not a fruit tree on the place. The catalogues of the nurserymen are pondered over by the young couple every evening, and sometimes on Sundays. At length the trees are selected. The best part of the farm is set apart for the orchard. The ground is ploughed deep, thoroughly harrowed, and then staked out. The rows are made scrupulously straight both ways. The holes are dug wide and deep, and partly filled with rich soil, or well prepared compost or garden mould. Then the trees are to be bought and paid for—fifty or one hundred dollars cash. Next they are planted, and oh, what care is taken that this shall be done exactly right. The young wife often goes out to help in this labor of love. She holds the tree while her husband is down on his knees filling in the earth about the roots with his hands. She reads the name on the label—"Sweet Bough." Perhaps there is a baby in the house; and she says, "How the baby will enjoy these apples when they are ripe!" The next is a Spitzenberg. And she says, "How good these will be in the winter—the long evenings—*and the pies!*" This young woman's mother was probably from Esopus, and had taught her daughter to believe in Spitzenbergs for pies and apple dumplings. If the Spitzenberg was as good in all parts of our country as it is in the neighborhood of its native place, it would stand at the head of the list for such purposes. But in New England the Greening is the favorite, while in Eastern Pennsylvania the old-fashioned Pennock is very good—or, at least, it *was* very good.

This is the time of Promise. That orchard is most carefully cultivated year after year with potatoes and other crops that will not injure the trees. No grain is ever grown there. The ploughing is done with oxen, that there may be no whiffle-trees to injure the bark of the young trees. At length the trees begin to blossom. Blossoms are always pretty, but none have ever been so pretty as these. There is some young fruit, but it falls off. No matter; there will be the more next year. The trees now grow beautifully; how large, and what a dark green the leaves are! How often the orchard is talked about! Every visitor must go to see the orchard. It is the Central Park of that couple's little world.

Six, eight, and ten years have passed. Several babies—but still no "Sweet Boughs" for summer, no "Spitzenbergs" for winter. The Apple Tree Borers are sometimes found; they came from the nursery in the young trees, but have been dug out and destroyed. Some of the Pear trees have been killed with Blight. Bark lice have been troublesome, but have been subdued by proper washes. Tent caterpillars have come from the neighboring orchards, or the neglected hedge-rows of wild cherry trees; but the clusters of eggs left by the moths on the twigs have been cut off in the winter, or the caterpillars in their nests have been killed when too young to have done much mischief. Other enemies have been kept in subjection by this careful, pains-taking, industrious young farmer. *But the fruit all falls prematurely, and what shall be done?*

Reader, if the ground under your fruit trees presents the appearance in midsummer of the one on this Plate, either the Curculio or Apple Moth, or both, have been there. Find out for yourself what it is, by cutting into these young fruits, and contrasting the living things you find there with the grub of the Curculio on Plates 3, 4, and 5, or the larva of the Apple Moth on Plate 9, Figure 1. If the grub of the Curculio has been the cause of all this falling, you know what to do. Every fruit that is destroyed by that enemy falls to the ground with that young grub inside of it, and continues there long enough to give the fruit-grower who chooses to destroy it, ample time to do so. If, on the contrary, it is the larva of the Apple Moth, as it very often is, in the Apples and Pears, then the case is different. Many of these caterpillars, or "worms," as they are usually called, will escape from the fruit before that fruit comes to the ground. In that event you want some way of trapping them. It has been long known that these Apple Worms, as well as some other caterpillars, will take advantage of the protection of cloths, old fragments of leather, pieces of boards lying near together, &c. I have seen several notices from foreign papers recommending plans of this kind for the destruction of some species of caterpillars that occasionally appear in great numbers in Italy and France. Some of the agricultural papers of our own country also state that the Apple Worm can be caught in a similar way. Dr. Harris, in his work on Injurious Insects, mentions it; but I do not know that any one has ever put a plan of the kind in practice to an extent that would really test its value.

Two years ago I took from the crotch of a young Bartlett Pear tree in the orchard of my friend Dr. Ward, near this city, an old boot-leg that had been doubled up and forced into that crotch. It had become so hard and dry, and the growing tree had pressed it so closely, that it had to be cut to pieces to get it out. This was in April. That old boot-leg contained in its different folds sixteen of the worms of the Apple Moth, in their larva or caterpillar condition, all snugly tied up in their silken cocoons. When these cocoons were opened the worms would creep off just as they would have done when taken from apples or pears in the fall or summer before. Since then I have tried everything I could think of that would be likely to suit the fancy of these little caterpillars, having this instinctive impulse to seek out places for concealment. The details of these various experiments will be found in the subsequent diary.

The result has been, that the hay-rope band, as shown in this Plate, is not only the cheapest and most easy of application, but the best of all the contrivances that I have tried thus far. But some people will say: It will take a great deal of hay to go over a large orchard in this way, and hay is very dear now. I have had a long fight with the insect enemies. There has been a good deal of wear and tear of patience. Job was a patient man—he bore all those boils with commendable resignation. Abraham Lincoln has been a patient man. To have borne all he has from the rebels on one side, and all their friends on the other, without once saying " by the Eternal," is a manifestation of gentleness almost superhuman. I am patient. A man who has fought the Curculio for so many years, must be patient. But when I meet a man who counts the cost of a yard of hay-rope, when he sees the ground covered with worthless fruit under each of those trees he has worked at so long and so faithfully, and with no apples, no pears, and no fruit of any kind—why then I lose my patience, and say—no, I won't say what I would say. Reader, go with me through the following diary about the Apple Moth—and then conquer it.

May 26.—In an examination of a box of larvæ of the Apple Moth, collected from the bark of Apple and Pear trees during the last month (April), one had become a moth, the others were about half in the pupa cases and half in larvæ. This is proof that there is a period of many days, indeed several weeks, between the appearance of the first and last of these moths, the larvæ of which have lived through the winter. The moths and butterflies of some species seem to come all together. They will swarm for a brief interval, and then as suddenly disappear. Such could be taken in the blaze of lamps or torches, as recommended by some writers; but these contrivances could hardly be made available for this one. This irregularity in their appearance proves also that most of them are not ready for the fruits till they are larger than they are now; see Apples and Pears of this date in Pl. 2, under head of Curculio.

July 14.—To-day I have wrapped hay-ropes round several Apple and Pear trees in Mr. P.'s orchard, three coils on each—one foot and two feet from the ground—and some round the large limbs five and six feet up; I have also used leather (chamois skins), old carpet, and cloths.

July 14.—Bartlett Pears show signs of Apple Moth. A very few have fallen. The little brown dust is issuing from the blossom end, and the black decaying spot near is to be seen in some; in others, the slightly discolored depression. Should all the fruit containing the larvæ of this Moth fall to the ground before it escapes, it would be as easily managed as the Curculio, by the grazing process; but as some of the caterpillars leave the fruit while it is still on the tree, the indications differ.

July 17.—Spent two hours to-day under Mr. P.'s Apple trees, cutting into hundreds of the blighted apples lying on the ground. Any one who wishes to know about the youthful state of the Curculio and Apple Moth can find out about it in this way. In some of these apples of the very early kinds, both these enemies had escaped. In testing this matter under one such tree, where none of the fallen fruit had been disturbed, the Apple Moth larvæ had escaped in the proportion of sixty to forty that were still to be found. Most of these forty were full grown, and ready to leave. Under those branches of the same trees where the fruit had been picked up a few days before, the apples that had fallen since, *nearly all* still contained the larvæ. In the later kinds, and especially winter sorts, as Reinettes, Baldwins, Spitzenbergs, &c., scarcely any had escaped. My experience to-day was more comforting as to our ability to control this formidable enemy of the Apple and Pear than I had expected. The stock that eats the fallen fruit promptly as it falls, will destroy a vast number of this enemy as well as the Curculio. Many apples that are brought down by the Curculio will contain the other enemy also. Some will have in them two or even more larvæ of the Apple Moth, of different sizes, so that if the apple should still hang on the tree till the oldest one escapes, it will be likely to fall before the others do. In this case some will become food for the animals.

This Apple Moth enemy I found more numerous to-day in these early apples than the Curculio. I have heard Mr. Carpenter, of Westchester County, N. Y., say, that this was their enemy of the Apple crop, and not the Curculio; but probably they have them both in about equal numbers, as in most other parts of the country.

It is true, undoubtedly, that many apples that have been bored through by this insect hang on the tree till they ripen. We find such at the cider mill, and in the market in the winter. Sometimes they will keep till spring. But suppose that all that are brought to the ground by the falling of the fruit are at once destroyed by the grazing stock, the aggregate will be so much diminished that the remainder will be more at the mercy of our adjuncts—the birds and the hay-ropes.

In some few instances I have seen where this insect, in its larva condition, has been victimized by a parasite; but when fairly housed under the scale of bark it is so completely out of the reach of these enemies, that we cannot rely much on them to assist us.

July 18.—About one in ten of our Bartlett Pears show signs of the presence of the Apple Moth. *Nine-tenths* of Mr. P.'s apples have already fallen from this and the Curculio. Two years ago my own Bartletts were very full, and so large a portion, at this time in the season, showed the presence of this enemy, that I determined to take every such pear off the tree before any had escaped. As many as half a bushel of these half-grown pears were taken from each tree at a time, and a few at intervals afterwards, and having no pigs they were fed to the horses. Last year these Pear trees took a rest—scarcely any fruit. Even the Bartlett has to rest sometimes—but nearly every pear came to perfection and was immensely large. Without the destruction of the crop of enemies in the season of full fruit, I could hardly have expected any in the season of so few. This is the great advantage of having control of the insect enemies of the fruits—you save the thin crops; and these are often the only crops of value for market. Fruit is scarce; and then, too, the few of the thin crop, if perfect, are so fine. We sometimes have a season of such an abundance of nearly every kind of fruit that the insect enemies take all they want, and we hardly miss them; indeed, I can imagine that we may have been benefited by the thinning out; but such a thinning out the next year may be fatal. This, too, is one reason of our fruit trees taking on the bad habit of irregular bearing. A thin crop taken off entirely by the Curculio or Apple Moth before old enough to have exhausted the tree, will often cause it to bear profusely the next; when, if the thin crop had been brought to perfection it would to some extent have been exhausted, and thus guarded against this injurious superabundance of the next year.

Much has been said about thinning out. I confess I have never seen it done. While the fruit is small it does not look so thick; and then, too, we cannot tell what may happen—it is all left. My experience is, that more trees are propped up, or broken for want of props, than are judiciously thinned out at the proper time. Perhaps the Curculio, Apple Moth, and other enemies were made on purpose for thinning out, and sent as a punishment for such shocking bad management as is shown in the breaking down of fruit trees from being overloaded. Even if you prefer to prop, let me beg of you to thin out your crop to the extent of taking off, before the middle of July, every pear that shows the signs of this little miner. That pear will be only a wind-fall at best; an insipid bite on one side, and the worm that caused it will have escaped to produce a whole brood of tormentors the next year.

July 26.—Have examined some Bartlett Pears that had fallen within the last three days from the effect of the Apple Moth. Found none of the larvæ—all had left. This was not the case ten days ago. Then many contained them; now, the safest plan would be to hand-pick from the tree.

July 28.—A few Bartletts fall every day; but the worm has now always escaped. Stock to eat this fruit would be useless.

Aug. 2.—To-day I have uncoiled one of the hay-ropes and found snugly concealed in their cocoons twenty-four of these little Apple Moth caterpillars. Some few of the caterpillars now so plenty had also chosen the dark recesses under these hay-bands as places of refuge. Spiders were there also. This is to be investigated further. If none of these Apple enemies have taken refuge in the scales of bark either above or below this trap, it augurs well for the success of this simple management.

Aug. 3.—I am now on a trip of observation to Western New York, and passing a night at Elmira. In a walk to the encampment of Rebel prisoners this evening, I saw many Apple trees in the gardens and grounds about some of the houses. The Apple Moth had laid the young fruit on the ground by thousands. Some trees near the doors of handsome mansions had shed so many that the ground was nearly covered with them. Nothing was done—the fruit never picked up—both Curculio and Apple Worm having their own way. The pigs were penned up, and were squealing for food. They would have enjoyed this young wormy fruit, but they squealed in vain.

Aug. 5.—CROOKED LAKE, YATES CO.—Plenty of marks of the Curculio on the apples about here, and still more of the Apple Moth.

Aug. 7.—NIAGARA FALLS—GOAT ISLAND.—A few Apple trees in a garden near the bridge. A full crop, but it will soon fall. Terribly infested with both Curculio and Apple Moth.

Aug. 10.—Home again. Have been examining the fruit in the barrels. I had placed, some weeks ago, a bushel of blighted apples in two old flour barrels half filled with earth and covered with millinet, but during my absence this covering has been disarranged, and the young enemies have mostly escaped. The Apple Moth larvæ were found in great numbers wherever they could conceal themselves about the old barrels. I observed the cocoons of six touching each other in a place where the hoop was about an eighth of an inch from the stave. Some fifty were found about the barrels in their larva condition, and three pupa cases. From this time I shall try to test what proportion undergo their transformation the present season. Some certainly do, and these Moths are probably the parents of the worms we find in the Apples and Pears so late in the summer and fall. Strange that there should be two broods in a season of some and only one of others. This is not the only instance, but such irregularities are rare. Bartlett Pears are still falling from the effects of the Apple Moth, but these wind-falls will soon be all off. The crop is somewhat thinned, but plenty left, and very fine.

Aug. 11.—I have at last found time to examine more carefully my Apple Moth traps set on the 14th of last month. On a young Bartlett Pear tree in my own garden, five inches through, I had wrapped a piece of Chamois leather, eighteen inches from the ground. The leather was folded into three thicknesses, and went twice round. It was secured by tying it firmly with twine near the upper part. The worms in ascending could easily enter under this leather, but could not get in from above on account of the twine acting as a ligature. Here I found twenty-one of these worms, all snugly wrapped up in their cocoons, and ready for their long sleep.

Some were between folds of the leather, some far in the crevices, and others in the fissures of the bark; the leather in this case answering the same purpose as the scales or layers of bark on Apple and Pear trees, and of course affording more protection from birds. Two feet higher up I had wrapped three coils of hay-rope, and there I found five more.

Aug. 12.—At Mine Hill, Morris Co., N. J. The country improved by the rains—all green; pasture good; none of the brown, parched appearance of Western New York a week ago. The heat, though severe, is not at all so oppressive as in the cities. Here I can rest; sleep comes again; can sleep in the woods or under the Apple trees. The heated brain is cooling off; the tension relaxing; mind is returning, and I am beginning to think. Apples fall around me, not at regular intervals, but often, day and night. A breeze rattles them down—all—all wind-falls. Attraction of gravitation brings them to the ground, but the Apple Moth gives gravitation the chance. Sleep again, and dream about Newton, the Principia, and Fruit enemies. Wake up; cut apples till knife and hands are black and sticky; find no grubs of the Curculio, but hundreds and thousands of their marks. From some cause or other, *they* have come to an early end—generally before they had destroyed the vitality of the apple. The Codling Moth enemy is the present cause of the most of this dropping. In cutting into these apples I often find one of the caterpillars, plump, full grown, and pink-colored, but most of them have already escaped; their excavations like black and mouldy caverns, the little pellets of their castings tied up together with silken cor's, and often stowed away in some deserted part.

As this caterpillar approaches full growth it makes an opening, generally through one side of the Apple or Pear, sometimes near the stem, occasionally at the blossom end, and there will be collections of the drillings pushed out, often looking like the chips in the side of a gimlet after boring unseasoned hemlock wood. This hole will sometimes remain plugged up for a time with these borings, and if you make an examination then, the worm will be found; but soon it will push out this plug and escape, whether the fruit is on the tree or on the ground. It has now come to the end of the eating period of life. In those Apples and Pears that ripen early, they mostly fall before the worm is grown, and it would then become the victim of the domestic animals, if these animals could have the chance; but at this time of the season most of the fruit hangs on till after the worm has escaped.

When it leaves the fruit, whether by day or night; and how it comes down the tree, whether by a cord or by creeping; it is hard to know. The next business is to find a suitable place of concealment from its enemies, and there to form its little cocoon. An old flour barrel will be used freely. The openings between the hoops and staves; the cracks between the staves that do not fit close; little pieces of split sticks tied together as faggots; two pieces of boards placed together and laid on the ground, or stood up endwise under or near the trees, will all attract them, especially if those trees are so young as not yet to have rough, scaly bark.

Some caterpillars are very particular as to where they attach their cocoons, but this one seems to have little choice. Wormy fruit carried into the house will prove this; the escaping caterpillars will find places to suit them in your furniture, your books, old papers, your clothing. I try so many experiments during the summer in my insect investigations, that I forget some of them. I have filled my pockets to repletion with the wind-falls of the little early pears, such as Doyenne d'été, Madeleine, etc. This coat has sometimes been hung up in a wardrobe, and the pears forgotten until they were *too soft*, and the worms have escaped. There will be slight obstructions in putting that coat on the next time. The sleeves will be found tied in places with little cocoons; pockets contracted. Where it hung in folds, will be little spots of flossy silk; and when these various cocoons are broken up, as they will be when it is put on, that coat will be found in a very wormy condition; and then if the hands are thrust into the pockets before remembering the summer pears—what a muss!

But in large orchards of older Pear and Apple trees, the rough bark seems the chief resort of this little insect. The layers or scales become the homes of nearly all. By placing these worms on such trees, as I have often done, and watching their habits, the instinct or reason, or whatever it may be called, becomes a matter of interest. How often they will peep into small crevices, and then out again! Sometimes they will creep in and come out at the opposite side, soon satisfied that they will not suit; sometimes remaining many minutes, and then looking further. When the proper place is at length found, little cavities will be dug out in the adjoining layers (Plate 10, Figure 1), and a firm border of silk will be made, inclosing these two cavities, and tying the scale to the true bark, and this often preventing those scales from falling off, as they would be likely to do by the warping liable to be occasioned by sudden changes of weather.

Aug. 20.—Many of the first crop of the larvæ of the Apple Moth are now matured—become Moths. That question is settled; there are two generations of this destructive pest in the same year. The Moths of the early summer are not those of this generation; these live but a few days. The twenty-one larvæ I took from under a leather round a Bartlett Pear tree some days ago, and which I placed in a large paper box, have all formed strongly-made cocoons. Three small pieces of the staves of an old flour barrel were put into the box, lying one on the other; they are now all tied together firmly.

Aug. 23.—I have to-day been examining my various hay-rope traps that were applied on the 14th of last month. I find that they should have been attended to sooner. About one in five of the worms have gone through their transformation and become Moths, leaving only the empty pupa case. Ninety-seven had taken refuge under one of these hay-ropes on an Apple tree (Pl. 11); forty-two under another; twenty-seven under another, and six under one that had been applied to a single branch of a tree. When three coils were made of this rope the worms were mostly found between the first and second coil, counting from the ground. They do not secrete themselves *in* the hay, but *under* it. If there are scales of bark on the tree, some use them also as an additional covering; but most of these had dug out little excavations, saucer-shaped cavities, in the bark, round the edges of which they had made their silken cocoons, and this cocoon lies in the dark protected space between the little concavity and the hay-rope, bound to the bark on one side, but not usually having any connexion with the hay on the other. This contrivance seems peculiarly attractive to these worms. The tree was large, and old enough to have the body well covered with scales of bark, under which they usually conceal themselves; but upon a careful examination only one could be found either above or below that was not immediately under the rope. These experiments so far are satisfactory. Leather, old clothes, and pieces of carpet, are all found to be attractive also, but in every instance where I have used either of these I have found more cocoons outside of them than where the hay has been applied. If the hay-rope will catch ninety-seven out of ninety-eight that take to the trees, it will prove an effectual way of getting rid of this most formidable enemy; and it is hardly possible to find any other material so cheap or so easy of application. In examining the trap, all that is necessary is to slip it up the body of the tree a few inches, and all the little cocoons, with the worms inside of them, are so perfectly exposed that nothing remains to be done but to crush them with the palm of the hand, either with or without gloves; then push the rope back again to the same place, or lower if necessary to make it as tight as it will well bear without breaking. One rope will last the season if carefully managed.

Pupa cases of the White Moth were found here. In opening one a swarm of two or three hundred little Ichneumon Flies came out. Others full of the maggots of these little parasites were found. These correspond in size and appearance to the *Microgaster* of Entomologists. They have all the characteristics of the Ichneumon class; the four wings, long, restless antennæ, and the constant motion. How wonderful these transmutations in the insect world! First the egg, then the caterpillar, then the pupa—then should be that beautiful white moth, but instead out comes this swarm of little flies. No wonder there have been people who believed in the transmigration of souls.

My friend Mr. P., who owns this fruit establishment where I pass so much of my time, was with me to-day when I was examining these hay-ropes, and watching the experiments with much interest; and when they were all counted where ninety-seven were found on the one tree, he remarked, "That will do—I can now save my apples. Instead of eight or ten to a tree I can have as many bushels." He has now promised to change his mind upon the subject of his insect enemies; and next year he will follow my directions and save his fruit.

Aug. 30.—Examined some of the hay-ropes again to-day. Under the one where I had found ninety-seven a few days ago, there were now ten more. These I brought home and placed on the ground near the foot of a Pear tree bound with a leather trap one and a half feet up. Most of them started at once to this tree, and were in a few minutes secreted under the leather. Three seemed to have lost the power of locomotion, remaining on the ground, where they were soon found and devoured by a community of small ants living in that neighborhood. So it goes—the destroyers of the apples were destroyed by ants.

Sept. 7.—I have taken to-day eight more Apple Worms from that tree in Mr. P.'s orchard, and have searched the tree faithfully both above and below the hay-rope, but find none except under the rope. These eight I placed near the root of the Bartlett Pear tree. They ran a little wild at first, but were soon all directed towards the tree, and in fifteen minutes were snugly secreted under the sheep-skin eighteen inches up. This kind of trap can be made effectual in subduing this formidable enemy. More experience seems to be only accumulative of what is fully proved. Whether it is better than anything else will require more time to ascertain; but if other contrivances should be found superior to this, the worms will certainly have but a poor chance to escape when they come to be generally used.

Sept. 8.—Visited orchards in the neighborhood of Clinton, Hunterdon Co., N. J. Find the Apples badly marked by both Curculio and Apple Moth. The bird enemy of the latter has been here too. Plenty of holes where it has pecked through. Fruit cultivation but little attended to in this neighborhood except the Peach, and the management of that is indifferent. The soil is suitable, and the crops would be good if better managed.

Sept. 13.—At the American Pomological Convention, Rochester, N. Y.; representatives from all the Loyal States, and some from Canada.

The fruits on the tables were fine, and in great variety—Pears, Apples, Plums, and Grapes.

By one who has not cultivated a propensity to seek out the blemishes in the fruits caused by the insect enemies, most of those on exhibition here would have been pronounced fine; but upon a close examination by eyes like mine, where the focus is fixed on defects, of the hundreds of plates, not one in twenty contained fruits that were all sound. Three-fourths had Curculio marks, and one-half had been more or less tampered with by the Apple Moth.

I was speaking on this subject to a company of delegates from different sections of country, when one from Wayne Co., N. Y., pointed to some of his on a table near us, as all sound. This challenged an examination, when more than half were found specked by one or the other, or both of these enemies.

If these specimen fruits were so defective, showed such evidence of the enemies, we may infer that those left at home, from which these were selected, were much worse; showing unmistakably that the time is rapidly coming when something must be done, or all will perish when there is a thin crop to begin with, and most that are left of the plentiful crops will be seriously blemished.

But the mischief has now become so manifest that I found it easy to produce a decided impression upon the delegates by some remarks I was called upon to make, and many expressed themselves under great obligations for the information I had given them.

Sept. 20.—Examined my worm traps again to-day. My previous accounts have been chiefly of one tree, and on that one I found twenty-four more. This is an Apple tree thirty inches in diameter, about twenty-three years old, and this year having at first quite a full crop. It had been several days since my last visit, and many of these worms had changed considerably in appearance, but the most of them were of the bright red or pink color, and active when exposed.

I examined several neighboring trees where no hay-traps had been applied, and could find plenty of these worms, but none on the others except under the hay. My good opinion of this mode of controlling this enemy is becoming more and more confirmed; and although new, I shall not hesitate to advise its general use.

Sept. 22.—Was at Trenton to-day. The fruit in market here is more disfigured by the enemies than I have seen it anywhere else this season.

Sept. 22.—Visited the old orchard, and found plenty of apple worms under the scales of bark. Did all the apples containing the larvæ of the Apple Moth fall to the ground while they are yet in them, as is so generally the case with the young Curculio, they would have been destroyed by this herd of cows, and I should have been unable to find them under the bark on the bodies of the trees.

Sept. 26.—Was at the rooms of the Farmers' Club in New York to-day, but the Horticultural Exhibition interfered with the meeting. The display of fruits was good, in some respects superior to that at Rochester. The Pears were very fine, and some collections little injured by the insect enemies, but the Apples were much specked. I examined carefully twenty-six plates of the latter, with five and six specimens on each, and only eight were perfectly sound. Not one plate with all perfect. The Apple Moth had done the greatest injury, but the Curculio had been meddling with a great many of them. Some of the collections of Pears had also been seriously injured. Even Quinces showed marks of both enemies.

Sept. 28.—Made another examination of the same worm trap, but found only two. The run is evidently nearly over.

Sept. 29.—Visited Easton, Pennsylvania, to attend the exhibition of the State Agricultural Society. The weather was very wet and disagreeable. Saw nothing except the fruit. The Curculio and Apple Moth had made their marks on this Pennsylvania fruit even more abundantly than on the New York fruits, at the American Institute in the city, or at the Pomological Convention at Rochester. Nearly all the Apples on the tables looked like wind-falls, and ripened prematurely. I sometimes visit the markets in Philadelphia, in the winter, and for some years past have seen more New York (or western) apples there than of their own native kinds. This was not formerly so.

Oct. 1.—Mr. Freeman, of South Orange, in this County, five miles from Newark, told me to-day that common apples were selling for 30 and 40 cents per bushel. Many have been sold as high as 60 cents for Harrison and Canfield; and the cider made from these two kinds finds a ready sale at $9 00 a barrel. He told me also that most of the cider-makers would not buy them, because "they were so generally bored by worms as to be dry, making but little cider." This is another phase of the Apple Moth question. Newark cider has been celebrated for many years, and commands a very high price. The best is made of a mixture of the Canfield and Harrison apples, and many farmers in Eastern New Jersey have large orchards, almost exclusively of these two kinds, but the Borers, Tent Caterpillars, Curculio, and Apple Moth have become such formidable enemies that the business is nearly given up, and must soon be abandoned altogether, unless these insect pests are resolutely met and conquered.

Oct. 6.—Have again examined my hay-trap on the tree at Mr. P.'s, but found only one worm. I have looked carefully again to-day, both above and below this trap, and find one more outsider. I have caught nearly 200 under the hay-band on this tree, and only two outside of it. A further examination on neighboring trees shows a great number of these worms in their usual winter quarters under the bark. From these experiments it looks as if one of these hay-traps to a tree would take all, or so nearly all, that but few would be left for the birds.

Oct. 16.—Passed two hours to-day in an orchard, a few miles from the city, examining the Apple and Pear trees. I found great numbers of the larvæ of the Apple Moth snug and close in their winter quarters. It is easy to see also where many more have been equally snug in other years, but had been found by their bird enemy. I observe no signs yet that these birds have commenced their searches this fall. Fewer holes are made by these birds near the ground than higher up. I find a few fresh-made holes on some of the Apple trees here, made by the Sapsucker.

Oct. 30.—The Apples in the New York market are now fine—many of them what they should be, perfectly sound. They are chiefly from Western New York, and bear strong marks of having come from the counties bordering on Lake Ontario; and any one accustomed to watching closely the fruit from that section will soon be able to distinguish it. But with all their perfections there are still enough blemished to tell that both Curculio and Apple Moth are there. Should all the apple-growers in that section of Western New York resolutely determine to conquer these enemies, "they would have the means of accumulating wealth beyond the dreams of avarice," as Johnson says.

Nov. 18.—Spent an hour to-day in examining the bodies of two old dwarf Pear trees in Mr. P.'s garden, near his asparagus bed, and found sixty of the asparagus beetles under the same kind of scales of bark where the larvæ of the Apple Moth are found—where lady-bugs, bouncing beetles, flies, and spiders are found. The day was mild, and all these were only semi-torpid. These asparagus beetles in several instances were snugly occupying the deserted cocoons made by the caterpillars of the Apple Moth. Possibly the hay-rope trap would suit the fancy of these pests. Will certainly try next year. I found about twenty of the Apple Worms on these two trees.

www.ingramcontent.com/pod-product-compliance
Lightning Source LLC
Chambersburg PA
CBHW030342170426
43202CB00010B/1215